TERRA PRETA

HAIKO PIEPLOW, HANS-PETER SCHMIDT
& KATHLEEN DRAPER

Foreword by Tim Flannery

TERRA PRETA

How the World's Most Fertile
Soil Can Help Reverse Climate Change
and Reduce World Hunger

WITH INSTRUCTIONS ON
HOW TO MAKE THIS SOIL AT HOME

DAVID SUZUKI INSTITUTE

GREYSTONE BOOKS
Vancouver/Berkeley

16 17 18 19 20   5 4 3 2 1

Greystone Books Ltd.
www.greystonebooks.com

David Suzuki Institute
219–2211 West 4th Avenue
Vancouver BC Canada V6K 4S2

Cataloguing data available from Library and Archives Canada
ISBN 978-1-77164-110-4 (pbk.)
ISBN 978-1-77164-111-1 (epub)

Editing by Shirarose Wilensky
Proofreading by Stefania Alexandru
Cover and text design by Nayeli Jimenez
Cover photograph by iStockphoto.com
Illustrations by Brian Tong
Printed and bound in Canada by Friesens
Distributed in the U.S. by Publishers Group West

We gratefully acknowledge the financial support of the Canada
Council for the Arts, the British Columbia Arts Council, the
Province of British Columbia through the Book Publishing Tax
Credit, and the Government of Canada through the Canada
Book Fund for our publishing activities.

Greystone Books is committed to reducing the consumption
of old-growth forests in the books it publishes. This book is one
step towards that goal.

CONTENTS

PART I: THE GROUNDWORK

Chapter 1: The Mistakes of Fossil Agriculture   5

The agro-industrial complex, often referred to as "Big Ag," is literally skinning the planet. Pesticides contaminate the ground; large-scale technology and modern chemical agricultural practices are causing the loss of the fertile humus layers at an alarming rate. This is one of the greatest environmental problems of our time.

Chapter 2: Cultures Need Fertile Ground— The Secret of Black Gold   33

Advanced cultures need a surplus of food that can only be harvested from fertile soils. This was the case 2,000 years ago in the Amazon region, where huge garden cities flourished on man-made terra preta soils.

PART II: THE HANDBOOK

Chapter 3: Climate Gardening—The Basic Principles and Materials
Through the use of climate gardening, huge amounts of carbon can be permanently stored in the fertile humus layer. Biochar and the stimulation of an active soil life are indispensable.

Chapter 4: Ways of Producing Terra Preta
Whether in compost heaps, kitchen bokashis, or slatted wooden crates—there are many ways of producing black earth. This chapter addresses many frequently asked questions and offers practical tips about the production of terra preta.

Chapter 5: Biological and Horticultural Diversity
Whether in urban or rural settings, private or communal gardens, window boxes or square-foot gardens, homemade terra preta is teeming with life.

Chapter 6: Old and New Sanitary Systems
Terra preta enables you to upcycle your own excrement, thereby reconnecting nutrient cycles. Homo, humus, and humanitas—it is no accident that the three words share the same origin.

═ FOREWORD ═

by Tim Flannery

OVER THE PAST decade, human emissions of greenhouse gases have reflected the worst-case scenario. In 2014, they reached 40 gigatons of CO2. That is a very large amount. If you wanted to take just 4 gigatons of CO2 per year (over a 50-year period) out of the atmosphere by planting forests, you'd need to cover an area the size of Australia.

Sadly, there is already enough CO2 in the atmosphere to push temperatures, in the next few decades, to 1.5 degrees Celsius (34.7 degrees Fahrenheit) above the preindustrial average. The lost decade of runaway emissions means that we will then be living in a different world. Australia's Great Barrier Reef, for example, cannot survive such warming. And, of course, we will not stop emitting greenhouse gases for some decades yet. It seems inevitable that we will emit enough to drive temperatures through the 2 degrees Celsius (35.6 degrees Fahrenheit) "safety rail" that scientists warn of.

These facts have profound yet poorly understood implications for the way we deal with climate change. Reducing emissions of greenhouse gases is critically important. But it is no longer enough. We need to find ways to draw gigatons of CO_2 out of the atmosphere. In *Atmosphere of Hope*, I argue that we can do this by fostering "third way" technologies. And few such technologies have greater potential than the production of biochar, the key component of terra preta. As you will discover reading *Terra Preta*, biochar is an old technology. But authors Ute Scheub, Haiko Pieplow, Hans-Peter Schmidt, and Kathleen Draper show that advances in production methods, combined with a fuller understanding of ancient techniques, have unlocked a very powerful tool for dealing with climate change.

By 2050, people will be living with a rapidly changing climate that will make feeding the expected population of around 10 billion a huge challenge. Biochar offers not only a way to reduce the climate problem but also to increase food security. I have no doubt that the biochar industry can grow to be a major contributor in the battle to stabilize our climate. Indeed, it must if we are to secure our future. *Terra Preta* represents a major advance in communicating how critically important biochar is. I hope that you are inspired by this book to become involved.

TIM FLANNERY

Create Your Own Paradise

WHEN PEOPLE TRY to imagine paradise, they usually think of a garden. We feel comfortable in gardens and recover from all kinds of stress. Gardens are places of richness, of fertility, and of plenty. We delight in them with all our senses—the warmth of the sun, the colors of butterflies, the shapes and scents of flowers, the smell of damp soil, the taste of fresh fruits, the sounds of rippling water and courting birds—or simply enjoy the peace and quiet. Maybe, now and then, a fat slug slithers through the idyll, and you imagine it chuckling to itself because it has just feasted on a fresh young lettuce that you had lovingly tended—something in paradise has to be annoying; otherwise, it would be boring.

Green is healthy and beneficial: people of all cultures and nationalities prefer a natural landscape to a concrete urban environment. Studies have shown that in hospitals a view of greenery from a patient's window encourages the healing process and reduces the dosage of painkillers that would otherwise be required. Trials in Germany, Holland, and the USA have demonstrated that nature in the direct vicinity has a positive effect on the contentment and health of those living and working there. Gardening is all the more healthy, as you are moving around in fresh air, light, and sun the whole day. And in the evening, you can reflect with satisfaction on your day's work—if only it wasn't for the pesky old slugs.

So it's not surprising that not only in the countryside but also in cities gardening is increasingly in fashion. Young and old; women, men, and children; those with or without employment; natives and refugees work the land, form intercultural gardening associations, cultivate green plots behind their houses, or tend some piece of communal land. Some people also do it because they are fed up with the callous production of food and the perverted globalized power of the agro-industrial complex. It is above all to this group of mostly ecologically minded urban and hobby gardeners that this book caters.

The book's most important message is a hopeful one: climate gardeners using terra preta can reduce the greenhouse effects on the earth's atmosphere. Climate farming—applied globally and supported by other environmental measures—could even reverse them. If we, by means of climate farming, were to raise the humus content of soil by 10 percent within the next 50 years, the carbon dioxide concentrations in the atmosphere could potentially be reduced to preindustrial levels. Maybe, as the Australian professor in earth and life sciences Tim Flannery stated in his open letter on biochar, the use of biochar "represents a cornerstone of our future global sustainability."

Terra preta, which simply means "black earth" in Portuguese, is also sometimes termed "miracle earth." The latter is certainly an exaggeration; after all, it is just a man-made fertile soil and not a witch's brew that you throw over your left shoulder at a full moon, spit on the ground, and say abracadabra, and the world is saved. Nevertheless, this ancient cultivation method used by the original inhabitants of the Amazon region has the potential, according to a growing community of international scientists, to tackle the global problems of hunger, poverty, water shortage, and climate change simultaneously. Which other methods can make such a claim?

According to current knowledge, the advantages of terra preta techniques are immensely diverse—they encourage fertile soil, healthy plants, and food; enable high and reliable yields within a small area; relieve smallholders and gardeners from dependence on expensive fertilizers, toxic pesticides, or genetic technology; transform waste products into natural manure; solve problems of hygiene in slums and sanitary systems; detox soils; transform steppes and poor soils into agricultural land; and decisively lessen the effects of climate change.

How? The key component of terra preta is biochar, charcoal made from organic residues and produced by means of climate-neutral pyrolysis, exposure to heat in a low-oxygen environment. If you pyrolyze organic wastes, up to 50 percent of the carbon, which plants have extracted from the atmosphere in the form of carbon dioxide, is converted into highly stable carbon, which can persist in soils for thousands of years. When introduced to the soil, biochar, which is rich in pores, serves as a fertile store of nutrients and water. How this actually happens will be explained in this book. Every hobby gardener and every smallholder can produce their own version of terra preta; if you have too little time, you can also order the necessary ingredients via the Internet.

However, the details of what these methods can and cannot achieve have by no means all been researched. A hurdle to its large-scale application in agriculture, at least for the time being, is cost-effective production of large amounts of biochar by modern pyrolysis facilities.

Conventional agricultural research institutes also show little willingness to finance and participate in appropriate research projects. This is hardly surprising, as there is the potential in black earth techniques to replace, or at least reduce, agrochemistry and genetic technology, the purveyors of which fund an increasing amount of research at agricultural universities. Similar to decentralized renewable energy, which in the long term threatens the existence of the central energy corporations, the increasing spread of terra preta methods would ultimately cut into the profits of agro-industrial players. We have to be prepared that at some stage, those promoting biochar will feel their opposition, which may take the form of excessively high safety regulations that the lobbyists of the agro-industries may promote to increase barriers to entry in the agricultural sector.

THIS BOOK IS split into two parts. The first section deals with basic principles, and the second section moves towards practical applications. Those of you who are in a hurry can skim the first part, though, of course, we would recommend reading it; otherwise, it might be difficult to understand the full potential of terra preta.

The first chapter of section one is about the mistakes of fossil agriculture. In all probability, we have passed not only "peak oil"—the point in time when the maximum rate of oil extraction has been reached and after which easily extractable oil amounts shrink and become increasingly expensive—but also "peak soil"—the time when the amount of non-toxified fertile land is decreasing and becoming ever more costly. Unlike climate issues, however, the soil has, up to now, no political lobby, though its global state and the impending climate catastrophe are closely linked.

In the second chapter, we make a short excursion through human history around the globe. We see how fertile land and advanced cultures are mutually dependent and how, in the past, whole empires disappeared because of soil erosion. We also illustrate the opposite: terra preta enabled huge garden cities to flourish in the pre-colonial Amazon, and man-made black earth has in the meantime been discovered in an ever-increasing number of indigenous cultures around the world.

The second section, the practice of using terra preta, begins with chapter 3. We describe climate gardening and climate farming (sometimes called carbon farming), practices that seek to offset the effects of global warming. How do they work, what are the basic materials and core principles, what effect are they having and where are they already in operation?

In the fourth chapter, we deal with the various methods of terra preta production, from compost heaps in small gardens and kitchen bokashis to stacked compost crates on balconies or in communal gardens. Additionally, we answer questions arising from the practice of urban gardening and climate gardening.

In the fifth chapter, we show the horticultural diversity in which terra preta finds its organic place. We describe how simple and attractive measures such as window boxes or tree patronage can reanimate the seriously threatened diversity of species in urban and rural settings and take readers to the impressive world of permaculture (highly efficient, self-sustaining systems that integrate human activity with nature) and community gardens.

In the sixth chapter, we deal with old and new sanitary systems that help restore nutrients from human excrement to the natural cycle. Here we come up against acquired taboos but demonstrate, for those who want to, how to realign our inner values with the cycle of nature. Just to make things clear: this isn't a must for the production of terra preta. Terra preta's key component is biochar, not human excrement. Those of you who feel put off by this topic, which as we will show there is no need to be, can skip this chapter.

In each chapter, you will also find short portraits of flourishing terra preta and climate farming projects.

Just one final comment. As it is barely more than a decade since the rediscovery of this ancient horticultural technique, the practical horticultural experiences are unavoidably limited. There is still a huge amount of details to research, such as which kinds of vegetables prefer which kind of substrate and what are the ingredients of this substrate. The more gardeners gather and share their experiences, the quicker knowledge about terra preta will spread. Let yourself become infected by the terra preta bug and talk about your experiences and observations with your neighbors, or via social media, local journals, or the radio. You may also join the new *Biochar Journal* and exchange your newly gained knowledge with other climate gardeners and farmers.

When the first edition of this book was published in 2013 in Germany, many readers began experimenting with terra preta recipes in gardens, on farms, or on urban terraces. It was only three short years ago, but since then, plenty of new hands-on experience and additional scientific results have been collected, and successful examples have multiplied all over the world. Perhaps the most important advance might be that farmers and gardeners are now able to produce at low costs their own high-quality biochar, one of the most important ingredients of terra preta, instead of buying it from industrial producers. To provide the most up-to-date information in this do-it-yourself guide, the English edition of the book was augmented to include new compelling climate farming stories, improved recipes for soil fertility, and instructions for how to make biochar. Replies to additional frequently asked questions have also been incorporated.

How long the soil of our one and only planet remains fertile is in our hands. With terra preta and the knowledge of how best to use it in different scenarios, we can begin to create a modern garden paradise worldwide where we can live better than we do now: a paradise 2.0.

BERLIN/VALAIS/ROCHESTER, FALL 2015
UTE SCHEUB, HAIKO PIEPLOW, HANS-PETER SCHMIDT,
KATHLEEN DRAPER

PART I

THE GROUNDWORK

CHAPTER 1

The Mistakes of Fossil Agriculture

The agro-industrial complex, often referred to as "Big Ag," is literally skinning the planet. Pesticides contaminate the ground; large-scale technology and modern chemical agricultural practices are causing the loss of the fertile humus layers at an alarming rate. This is one of the greatest environmental problems of our time.

THE DIAMETER OF the earth is a good 12,000 kilometers (7,900 miles); measured against this, the humus layer of 5 to 50 centimeters (2 to 20 inches) is microscopically thin. Almost the whole of terrestrial life is dependent on this delicate, vulnerable layer. It allows plants to grow, which in turn feed people as well as supply oxygen. Without these complex intercouplings of life cycles, Earth would be as desolate and empty as it was in the beginning.

Set the book aside for a moment and go to the nearest piece of land—beneath a tree on the street or a plot of grass outside your house—and scoop up a handful of earth and smell it. Is it sandy or solid and compact; does it smell neutral or musty? Unfortunately, that's not all that uncommon nowadays. The thin skin of the earth is as good as dead in many places, and that is reflected in how it smells: lifeless.

If the earth is dark, warm, loose, and crumbly and smells like a forest floor, then you are in luck, or as an enthusiastic organic gardener or passionate soil-lover, you are sitting in the middle of your own garden. The handful of healthy soil you now may be holding in your hands consists of the organic remains of dead plants, fauna, and microbes, small stones, minerals, water, soil vapor, and plant roots. Soil rich in humus is not called Mother Earth for nothing, and every small sample contains an infinite variety of life. The number of the microscopic organisms in the healthy soil you may be fortunate enough to be holding in your hand is many times greater than the entire population of humans on Earth.

AN INCREDIBLE ABUNDANCE OF LIVING CREATURES

IN THE TOPMOST 30 centimeters (1 foot) of 1 square meter (11 square feet) of healthy soil there is an incredible abundance of living creatures: an estimated 1 trillion bacteria, 500 billion flagellates, 100 billion rhizopods, 10 billion actinomycetes, 1 billion fungi, 1 million algae, 1 million paramecia, 1 million nematodes, 50,000 furculae, 25,000 rotifers, 10,000 polychaetes, 300 polypods, 150 insects, 100 dipteran larvae, 100 beetles and larvae, 80 earthworms, and 50 each of spiders, snails, and wood lice.

Microorganisms are the oldest form of life on Earth; incredible numbers also live in our bodies—10 times the number of our body's cells. We are inundated with them, but without them we could not live. They occupy all surfaces, they protect the skin from pathogens, and they live in our intestines and control our digestion. We are constantly accommodating new microorganisms and dispensing with others. They ensure that infectious organisms don't develop, and even pathogens have their function. When life expires, they create the prerequisites of a new beginning through decomposition.

But in nature, it is not decomposition that is the dominant process but rather generation, cooperation, coexistence, and symbiosis. Fungi, algae, and bacteria live in close relation to plant root hairs, mutually exchanging large molecules such as proteins, vitamins, or even whole cellular power sources such as mitochondria.

Researchers using soil-screening techniques have projected that up to now only 5 to 10 percent of all soil organisms are even known. With such limited knowledge, it is impossible to clearly know and recognize their diverse reciprocal effects, coexistences, and symbiotic relationships. The reuse of metabolic wastes is one of the secrets of living, fertile soil. We threaten this world with every dose of synthetic fertilizer, with every drop of toxin from pesticides, with every liter of slurry, and with every tillage of the soil with equipment weighing many tons.

Soil can indeed regenerate, but it needs a lot of humus and constant supplies of organic materials. If, however, the humus gradually disappears, then the soil dies and becomes stone and dust. With industrial farming waging a chemical and mechanical war against nature, soil can be quickly converted into what is commonly referred to as dirt, a lifeless substance incapable of producing living plants or hosting

HUMUS

ALL THE DEAD and decomposed organic substances in soil are collectively termed humus. It consists of a multitude of complex compounds that are transformed by soil organisms. Carbohydrates and proteins decompose quickly, whereas cellulose and wood components take longer. Humus, however, is far greater than the sum of its biological, chemical, or physical attributes; it is the indispensable basis for life in and on the ground and acts almost as an independent living organism. Plants provide soil organisms with vegetable saps and necrotic vegetable residues and receive in return nutrients—each living off and providing for the others. The late microbiologist and natural philosopher Raoul Heinrich Francé succinctly summarized this process: "Humus is made from life, by life, for life."

living animals. When that happens, we lose our footing and quite possibly our civilization as well.

The role of the earthworm in all these processes cannot be underestimated. In one acre of healthy garden soil more than 50,000 earthworms wriggle about, collectively weighing about as much as several cows. In its intestines and by using muscular power and countless microorganisms, the worm transforms dead vegetable matter and soil particles into high-grade soil—processing up to 70 times its own weight annually. In earthworm excrement, on average, you can find twice as much carbon, 5 times the amounts of nitrogen, and 7 times as much phosphorus as in normal soil. The tunnels a worm digs aerate the soil and provide growth pathways for plant roots.

Charles Darwin—a keen observer and great worm-lover—discovered they could even grind small stones to make mineral soil. "It is a marvelous reflection," he wrote in *Vegetable Mould and Earthworms,* "that the whole of the superficial mould over any such expanse has passed, and will again pass, every few years through the bodies of worms... It may be doubted whether there are many other animals which have played so important a part in the history of the world, as have these lowly organized creatures." Ultimately, all our food and thus, indirectly, we as human beings have all passed through the bodies of worms.

Another important role is played by mycorrhiza (Greek: *mykós,* "fungus" and *riza,* "root"), a symbiotic association between fungi and the roots of vascular plants, mutually unlocking important nutrients. Also indispensible are bacteria that capture nitrogen from the air and form symbiotic relations with leguminous plants, such as peas, beans, clover, and lupines. They are some of the very few organisms that can convert the nitrogen that is readily available in the air in a plant-available form and direct it to a plant's roots, which is why organic farmers plant "nitrogen gatherers" like the leguminous plants next to "nitrogen consumers."

Plant communities and soil organisms form highly complex ecological units. Plants communicate with each other using biochemical signals, giving reciprocal warnings of the approach of hungry insects. If they are attacked, plants can emit substances to repel certain creatures, or they can attract beneficial insects with other secretions. Beneficial insects can then make a meal of the plant's adversary. Using mixed cultivation, farmers and gardeners benefit from these symbiotic relationships in the form of higher and more stable yields. Mixed cultivation in forest gardens makes plowing unnecessary, which in turn encourages the development of humus. This explains why it is common to have high concentrations of humus in forests and permanent grasslands.

Humankind wouldn't exist without this thin layer of humus covering the planet, but it needs a great deal of time to become established. Natural soil formation rates vary from 1 to 2.5 centimeters (0.3 to 1 inch)—per century. If only 1 centimeter (0.3 inch) of soil is lost by human thoughtlessness, it can take 100 years to regenerate.

THE SOIL HAS NO LOBBY

NOWADAYS, FERTILE SOIL disappears 10 to 100 times quicker than it can regenerate. Since the end of World War II, around

1.2 billion hectares (3 billion acres) of arable land globally have literally gone with the wind—this is roughly equivalent to an area the size of India and China combined. Worldwide an estimated 24 billion tons of earth are lost to wind and water erosion—roughly 3 tons per person on the planet.

In this process, conventional industrial agriculture creates a fundamental problem: in the long term, it ravages the soil by promoting soil erosion at rates far exceeding the rate of soil regeneration. A team of scientists led by Professor Tom Dunne, from the Bren School of Environmental Science and Management (University of California, Santa Barbara), analyzed historical soils in Africa and concluded that the natural erosion rates amounted to about 2.5 centimeters (1 inch) in 900 to 3,000 years. Nowadays, these erosion rates are achieved in only 10 years.

The bottom line is we are treating soil like dirt. In the last 25 years, a quarter of the whole earth's surface has become degraded. The devastation and desolation of entire areas is proceeding at an unprecedented pace. In Spain, for instance, 40 percent of the soil is already damaged. In many regions, especially in Africa, the humus content is only about 1 percent, well below critical thresholds needed for growing crops. This is one of the primary reasons why more than a billion people go hungry; a heartrending fact considering we live in a world that is intrinsically rich.

Soil—one of the most important raw materials of all life is at the same time one of the most disregarded. The former German minister for the environment and executive director of the United Nations Environment Programme (UNEP) Klaus Töpfer criticized the fact that soil lacked a lobby at a press conference in the summer of 2012 and stated his intention in the coming years to establish an international network called Global Soil Week to publicize findings about damage to all earth strata. As if to prove his theory, a mere three journalists were present for a discussion among high-ranking experts, including the head of the UN offices for combating desertification. Had the discussion been about big business or the economy, the hall would have been packed. However, things have started to change in the last three years, as at least the media is giving more attention to soil. That 2015 was the International Year of the Soil helped somewhat.

It is generally not ordinary people who abuse the soil; it is the large agricultural corporations, otherwise known as "Big Ag," promoting their heavy machinery and chemical cocktails, that carry most of the blame for skinning the planet. Fossil fuel–based agriculture is a key contributor to the closely linked crises of climate, humus, and hunger.

We shouldn't forget that only 160 years ago, feeding the human population was based exclusively on nonchemical farming. The first settled populations in the Neolithic and Bronze Ages turned forests into arable land using slash-and-burn techniques and used their fields and pastures until they were no longer productive. Afterwards, they progressed to using two-field and, later still, three-field crop rotation, leaving fallow fields to regenerate. Through the ages and in all corners of the world, groups of farmers used the knowledge that both animal and human excrement restores fertility to the soil. In many agricultural societies it was, and in a much smaller number of communities still is, considered polite for guests to visit the outhouse after a feast to leave behind some nutrients. Together, plants, human beings, and domestic animals formed a nutrient cycle that continued until the beginning of the industrial age.

FOSSIL AGRO-INDUSTRY REACHES ITS LIMITS

AROUND 1840, THE German chemist Justus von Liebig discovered the growth-stimulating effects of nitrogen, phosphate, and potassium on plants. Mineral fertilizers replaced both animal manure and plant residue organic fertilizers. From the beginning of the twentieth century, first in America and then in Europe, tractors and haulers began to replace plows drawn by livestock. Fields were enlarged and large monocultures steadily began to emerge. Increasing amounts of fertilizers caused yields to expand rapidly to feed a growing world population.

Today, it is becoming ever more apparent that this system has reached its biological limits and is no longer sustainable. Synthetic fertilizers become more expensive and will decline, just like the fossil fuels for the tractors and factory farms. Heavy equipment compacts the soil and destroys soil life. Nitrogen, potassium salts, and pesticides contaminate fields and oceans. Monocultures accelerate the extinction of species and climate change. Genetically modified plants and hybrid varieties that have lost the capacity for reproduction threaten the food sovereignty and independence of farming communities and whole societies.

In 2009, a study by an international team of scientists led by Johan Rockström, published in the journal *Nature* under the title "A Safe Operating Space for Humanity," came to the shocking conclusion that the extent of the global loss of bio-diversity together with human interference with the nitrogen and phosphorous cycles was far more serious than climate change. In these three fields, the man-made problems are so great they threaten the very existence of humanity.

Big Ag is partially, if not directly, linked to these three fatal problems: loss of biodiversity, nitrogen and phosphorous cycles, and climate change:

- Today, animal and plant species are disappearing at least a thousand times more rapidly than in the last 60 million years. One-quarter of all mammals, almost a seventh of all bird species, and two-fifths of all amphibians are endangered. As specialized species lose their habitat and food becomes scarce because of agrarian monocultures, there is often more biodiversity to be found in urban areas than in the country-side. Strangely, we have already become used to the fact that many villages are no longer humming and buzzing rural gems but places that stink of slurry or landfill.

- The diversity of crop plants and domestic animals resulting from 10,000 years of agriculture is now rapidly declining, because agricultural corporations are only interested in cul-tivating, breeding, and monopolizing high-yield and hybrid species. Of the 8,000 known livestock breeds, above all, cattle, pigs, and chickens, 1,600 are endangered or extinct. According to estimates of the Food and Agriculture Orga-nization of the United Nations (FAO), in the last 100 years, three-quarters of crop varieties have vanished. Among these were probably a large number of species that would have been more resilient against the extremes of climate change than those used today.

- The global nitrogen cycle has been severely disrupted since the development of the Haber process more than 100 years ago. This process, which was patented in 1910 by BASF (nowadays the largest chemical producer in the world and based in Ludwigshafen, Germany) and was used exclusively for manufacturing explosives during WWI, uses fossil fuels to produce synthetic nitrogen fertilizers. Each year, 115 million tons of nitrogen are artificially produced. In the meantime, around 40 percent of the nitrogen in each of our body's cells contains nitrogen that was once produced industrially as fertilizer and subsequently taken up by plants that became either food or feed and was thus eventually assimilated by the human body. Nitrate leaches from overfertilized fields and ends up in rivers and oceans, which causes toxic algae to prosper and creates ever-widening dead zones. At least 10 million Europeans drink nitrate-polluted water. According to a study by the Centre for Ecology & Hydrology, Edinburgh, the nitrate contamination of the air, ground, and water causes costs for health care, ecosystem services, and water treatment that add up to between 150 and 740 euros per person annually in Europe.
- Phosphorus, essential for the survival of plants, animals, and humans, is a scare mineral resource, and peak phosphorus may be reached as soon as 2030. According to the German Federal Environment Agency, there are supplies for up to 80 years, but these supplies are increasingly contaminated by cadmium and uranium and will become less and less suitable for agricultural use. Already, an estimated 12,000 tons of uranium are unintentionally applied every year through naturally contaminated phosphorus fertilizer on the world's fields—no one has ever properly analyzed the effects on

reach the butcher's slab, as many greenhouse gases have been released as a 250-kilometer (150-mile) car trip.

- The amount of fossil fuel used for the production of one liter of conventional milk, for instance, would be sufficient to transport this same liter once around the world. The path of the milk production begins in Paraguay or Brazil, where genetically modified soybeans are grown for animal feed. The soy is shipped to a European harbor, often Amsterdam, where it may be sold on to dairy farmers in Russia and used in the production of powdered milk, among other things, and exported to, say, Kenya or Jamaica.

- The vast expanses of monoculture crops, without any sign of trees or hedges, contribute to the exacerbation of global warming, because they disrupt the natural evaporation cycles and prevent the earth from cooling down. Additionally, monoculture cropping endangers the earth's most outer layer, as wind and weather have free rein to erode fertile soils once the crop has been harvested. Scientists have estimated that soil is eroding at a rate 100 times faster than the ability to regenerate it, largely because of industrial agricultural practices such as plowing and fertilizer use. While these practices are causing the return of dust bowl–like conditions, they are being exacerbated in the western U.S. and elsewhere by a multi-year drought that has caused farmers to quickly deplete aquifers. Aquifer depletion is causing the land to sink, in some places significantly. This rapid subsidence is affecting infrastructure, including roads, bridges, and canals.

This form of fossil fuel–based farming, with its enormous waste of natural resources and destruction of natural habitats, leads to a very bleak future. Although the recent trend in fracking seems to have increased the supply of fossil fuels, at least temporarily, oil is irrevocably reaching its end. These fuels, however, come with a high environmental cost and consume massive amounts of freshwater, which is beginning to result in water wars between farmers and frackers in some areas. We cannot feed the 9 billion people expected to populate the planet by 2050 with this type of unsustainable agriculture. If we fail to change direction, we will live to see fierce and bloody battles over the allocation of scarce resources.

THE RETURN OF PLAGUES VIA MEAT CONSUMPTION

THE NATURE AND scale of industrial meat production, at the expense of developing countries, is raising many social justice questions. Roughly 26 percent of arable land on our planet is used for grazing livestock, with an additional 33 percent used to cultivate grain to feed livestock. In Brazil and Paraguay, forests are being cleared to plant genetically modified soybeans that are fed to cows and pigs raised in concentrated animal feeding operations (CAFOs) in China and other countries without sufficient land and water resources to grow enough grains to feed the growing population of livestock. Meat from these CAFOs ends up in the bellies of those who can afford to eat meat regularly but then may go on to cause cardiovascular disease, the treatment of which carries high economic costs in the form of health care. Almost everyone loses at this game, with the exception of Big Ag and possibly Big Pharma.

Another example can be found with broiler chickens, of which Europeans prefer to eat only the breast and fillets.

The leftover is shipped in refrigerated containers to developing countries, where it is sold for less than locally produced chicken, thereby wreaking economic havoc on local production, as 1 kilogram (2.2 pounds) of European poultry costs roughly 1.40 euros in Benin, whereas local meat is 2.10 euros. Chicken consumed in many parts of the world is often anything but harmless, as many chickens raised in large operations are fed a steady diet of antibiotics to prevent them from getting sick and facilitate weight gain. In fact, the World Health Organization (WHO) estimates animals consume three times more antibiotics than humans. Continual administration of medications results in multiresistant viruses that can no longer be controlled by antibiotics. This could increase health care costs to between $21 billion and $34 billion in the U.S. alone.

"You are what you eat," as the saying goes, but who says whether the effects of qualitatively inferior food are limited just to our bodies? Body and soul are inseparable and have a synergistic influence. Could the substances in industrially produced plants and animals from mass-breeding programs that provide much of our nutrition be responsible for the steep rise in human depression in industrial countries?

Cattle pens have long been breeding grounds for all manner of pathogens. As with broiler chickens, CAFO-raised cattle are routinely fed different types of antibiotics—creating ideal conditions for the cultivation of resistant germs. CAFO-raised animals easily infect each other, as these compounds house hundreds, or even thousands, of cattle in cramped conditions full of their own excrement. Antibiotics, bacteria, and pathogens such as *E. coli* and salmonella end up in the manure slurry and from there are often transported directly to the fields. Sometimes the toxic manure makes a detour through an anaerobic digester (AD), after which the fermentation residues, called digestate, are often applied to the fields. Bacteria can survive there, outside the intestines, for up to a year. Scientists demonstrated some 10 years ago that bacteria can be transferred in this manner.

E. coli contamination is not only expensive but can sometimes be deadly. The USDA's Economic Research Service estimates the annual cost resulting from E. coli to be nearly $480 million (73,480 infections, including 61 deaths, at an average cost per case of $6,510). In late 2015, Mexican grill chain Chipotle, known for using fresh, high-quality ingredients, was forced to close 43 restaurants after 45 customers contracted E. coli. The chain has announced additional infections in 6 different U.S. states. Although the source for this particular outbreak might not ever be known, retail wholesale giant Costco quickly found tainted celery produced by Taylor Pacific, the world's largest producer of fresh-cut vegetables in California, to be the source for 19 infections in 7 different states. Although the direct and proximate cause of the E. coli has yet to be identified, given that California is the leading dairy state in the U.S., with more than 1.7 million head of dairy, it is likely that the use of contaminated manure on fields may be connected to these repeated outbreaks.

Research out of McGill University in Montreal has shown that biochar-amended soils can reduce leaching of fecal coliforms into groundwater or neighboring water bodies. Additional research out of the USDA's Food Safety and Intervention Technologies Research Unit has shown that various types of biochars can inactivate E. coli in certain types of soils amended with 10 percent biochar by weight. Prevention of uptake and leaching are both critical to keeping *E. coli* out of the food chain.

GLYPHOSATE AND GENETIC TECHNOLOGY

MASS PRODUCTION OF meat is just one example of how fossil fuel–based agriculture will bring us to the proverbial dead end. Another huge area of concern to human health within agriculture is genetic technology. In a recent poll, nearly 90 percent of American voters would like to see GM foods labeled accordingly. Yet the U.S. Congress, flush with millions of dollars in campaign contributions from the industrial food complex, is working hard to block states from mandating GMO labeling. Fortunately, at least one state, Vermont, has prevailed and is set to require labeling by mid-2016.

Most people eat GM soybeans indirectly, by way of meat products. Almost 90 percent of animal fodder for European factory farming comes from imports, mostly from South America. There, soybean monocultures covering more than 40,000 square kilometers (15,000 square miles), an area as large as Germany and Switzerland combined, are expanding, two-thirds of which are GM soybeans.

Glyphosate is one of the most commonly used weed killers in the world. Yet it is suspected to be killing more than just weeds. Additives within glyphosate, such as tallow amines used as wetting agents, and its degradation product amino-methylphosphonic acid (AMPA) are suspected of disrupting cellular and embryonic development, as well as the hormone systems of animals and humans; promoting certain kinds of lymphatic cancers; and boosting the chances of developing kidney and liver damage. In Argentina, some 200 million liters (5 million gallons) of herbicides are sprayed annually, the most common of which is Roundup Ready, with glyphosate as its active ingredient.

According to studies by embryologist Andrés Carrasco, in the cultivation areas of GM soybeans in Argentina the rate of cancer among children tripled between 2000 and 2009, and the rate of miscarriages as well as deformities, some of them "freakish," quadrupled. Since the introduction of GM soybeans, Professor Damián Verzeñassi, from the National University of Rosario in Argentina, has also registered increases in brain malformations, hydrocephalus (water on the brain), cleft palates, and sironomeli—a deformity in which the legs are fused together, giving the appearance of a mermaid's tail.

Roundup Ready and other glyphosate preparations are deployed in vast quantities as perfectly legal herbicides in Europe. Companies such as Monsanto, Syngenta, Bayer, Nufarm, and Dow AgroSciences produce and distribute them. Half of the global production of glyphosate, however, comes from China. Untrained laborers spray these toxic weed killers on farms, garden plots, and even at schools and daycare centers in the quest to minimize weeds and maximize yields, without understanding the long-term negative impacts they are causing to humans. Searching for the causes of serious diseases in cattle, such as chronic bovine botulism, veterinarians and scientists repeatedly established the presence of glyphosates in the animals' urine, excrement, milk, and forage. A recent German study highlights some promising research done by veterinarians showing that feeding livestock a combination of charcoal, sauerkraut juice, and humic acid can help alleviate some of the negative health impacts suffered by animals fed glyphosate-contaminated GM grains.

At this point, glyphosate is omnipresent. In December 2011, the University of Leipzig, Germany, analyzed the substance in the urine of test subjects with readings varying between 0.5 to 2 nanograms per milliliter (ng/mL)—that is 5 to 20 times the limit values of 0.1 ng/mL for drinking water. None of the test subjects had direct contact with farming or herbicides, and many had lived for years exclusively on organic produce. Microbiologist Monika Krüger tested her colleagues and discovered that not one of them was free from glyphosate contamination. In September 2012, *Öko-Test* (a German consumer magazine) detected glyphosate residues in 14 of the 20 cereal products tested, including flour, bread rolls, and oats.

One cause of this high contamination is probably because of a process known as crop desiccation, which is sometimes used to facilitate harvesting of cereal crops, rapeseed, and pulses. Shortly before harvesting, glyphosate is sprayed directly on cultivated plants. Crops are thus killed off quickly and uniformly, allowing for an earlier and easier harvest, as the desired degree of dryness of cereals can be attained while at the same time weeds are destroyed before sowing the next crop. Despite all of their efforts, glyphosate probably reaches even the bodies of people on organic diets, either through occasional consumption of conventional cereal products or because of wind carrying the toxic herbicides to the neighboring fields of organic farmers.

MONSANTO

HARDLY ANY OTHER business has as many opponents as the U.S. company Monsanto, which was founded in 1901 and has subsidiaries in more than 61 countries. They used to produce the now banned toxins DDT and PCBS and supplied U.S. forces with Agent Orange, a defoliant that was used in the Vietnam War and caused serious deformities among the local population. Since the 1980s, Monsanto has been patenting animal genetics and cultivating genetically modified plants that are resistant to its agrotoxin, glyphosate. According to the International Assessment of Agricultural Knowledge, Science and Technology for Development (IAASTD), Monsanto is one of the 10 companies that dominate global food supplies. They control not only 90 percent of the genetically modified seeds but also 40 percent of conventionally produced corn and 25 percent of conventional soybeans. Excluding products from organic farming, Monsanto or its subsidiaries is responsible for every other cucumber, every third bean, and every fourth onion seed.

It boggles the mind to think that farmers do not understand or perhaps do not consider the downstream implications of spraying toxins onto their crops shortly before the cereals are harvested, threshed, and sold to bakeries. One might as well mix glyphosate directly into the dough.

SAVING THE WORLD WITH
SMALL-SCALE ORGANIC FARMING

WORLDWIDE, THE CONVICTION is growing that things cannot and must not continue as they are. The IAASTD's global report of 2008, *Agriculture at a Crossroads*, had a considerable role in the rethinking of agricultural strategies. It is the most comprehensive and informed analysis of current global farming practices. More than 500 experts from every continent and from all varieties of scientific disciplines worked on the report over a four-year period. Their message can be condensed into one sentence: continuing as we have been doing up to now just isn't possible.

The report goes on to state that a key solution to the crisis in global hunger and poverty lies in encouraging small-scale farming. These farmers still produce the majority of food, yet they are often susceptible to hunger and malnutrition themselves, as they typically cultivate only small parcels of land—most own less than two hectares (five acres)—and are heavily exposed to climate change, extreme weather events, or pests. These mostly organic family businesses should be properly supported, and their traditional knowledge of agricultural practices needs to be taken seriously; instead of being considered primitive, it needs to be encouraged and

understood, as very often these ancient practices are far better suited to local soil and weather conditions than generic
farming solutions advocated by Big Ag.

Another important message from the global report was
the widespread view that Big Ag and genetic engineering are
necessary for sufficient supplies for a growing world population, as propagated by their lobbyists, is nothing more than
a myth. Organic farming could completely cover the world's
nutritional needs, and according to studies, could supply
between 2,640 and 4,380 kilocalories per person per day.
Although scientific studies have shown that changing European and North American farming to organic production
would lead to a decline in total production and thus a drop in
exports, yields everywhere else in the world could increase,
even double, because in warmer countries the humus content of the soil is easier to increase when organic methods
are practiced. A much larger part of the population could
be fed on a diet of organically grown food if less meat was
consumed and food waste was reduced. A recent study from
the United Nations Environment Programme and the World
Resources Institute (WRI) found that nearly one-third of all
food produced worldwide, valued at US$1 trillion, ended up
lost, rotted, or otherwise unused.

FOOD SECURITY AND FOOD SOVEREIGNTY

IN THE COURSE of the twentieth century, we found ourselves on a path that began with genetic diversity but led to
homogeneity, as far as our cultivated plants are concerned.
From the countless species and varieties, such as amaranth,

quinoa, or millet, that were cultivated in various parts of the world, only 15 main crop species are commonly cultivated, with the bulk being: rice, wheat, corn, and soybeans. Rice (26 percent) and wheat (23 percent) are the top worldwide food supplies. Fewer and fewer varieties of these staples are being cultivated, with high-yield variants being preferred to regionally adapted versions. This loss of species diversity is a significant risk to global nutrition. Since 1960, 90 percent of all wheat varieties have disappeared, and the figures for rice and corn types are not much better, at 70 percent and 60 percent respectively. The other top crop species are cultivated in huge monocultures, which encourage the spread of plant diseases, harvest failures, and hunger.

The loss of species and varieties that suited local weather and soil conditions began in the 1930s with the cultivation of hybrids and spread rapidly after the Second World War. Hybrid plants use synthetic fertilizers more effectively, provide higher yields than their ancestors, and can be sown more densely. However, they can't pass on their attributes; consequently, new seeds have to be bought and planted every year. The farmers' prescribed dependency on seeds, fertilizers, and pesticides is intentional. Big Ag profits at the expense of small farmers around the globe.

With the loss of locally adapted varieties, knowledge about seed preparation and the culture of a variety of species becomes increasingly foreign to gardeners and smallholders. Thirty years ago, worldwide there were still more than 7,000 businesses dealing in the propagation of seeds, and none of them had a global market share of more than 1 percent. Today, three companies, Monsanto, Syngenta, and

Bayer, dominate two-thirds of the global market and control, to a large extent, which species are cultivated and sold. They also promote the distribution of genetically modified species, forcing the farmers to be ever more dependent. All of these chemical agriculture practices put our long-term food security at risk.

The high yields of GM "turboplants" are often planted in soils that must be heavily fertilized, as they have lost so much humus through overexploitation over the years. Indigenous crops that are locally adapted but offer smaller yields seem hopelessly overshadowed. However, recent trials have shown that farmers can attain high, and above all, safe yields without fertilizers, pesticides, and heavy machinery, even, in some cases, outperforming typical returns of monocultures when these regionally adapted varieties are cultivated on terra preta–enriched soils with a high humus component. Cultivation of one's own seeds would again be possible, as the harvested seeds pass on their attributes. Improving crop diversity and reducing their dependence on external inputs in the form of seeds and fertilizers could significantly improve farmers' incomes while providing a healthier diet to farming families.

Noncommercial seed banks also play an important role in saving the diversity of cultivated plants. In the 1970s, a virus destroyed nearly one-quarter of Asian rice production. Fortunately, a solution was found for this staple food for more than a billion people in a resistant wild rice species from a seed bank of the International Rice Research Institute in the Philippines. Bananas were in a similar situation. In genetic terms, banana plants are clones; they cannot reproduce sexually but depend on farmers taking offshoots from their roots to grow new plants. In the 1950s, a fungal pathogen known as Panama disease attacked the roots of banana trees; in short order, this disease wiped out nearly all banana plants in Central and South America. Thanks to a seed bank, a variety was found that was naturally resistant to this fungus—the now-ubiquitous Cavendish bananas.

In 1993, La Via Campesina, an international alliance of small farmers' organizations, was founded and coined the term "food sovereignty" as a counterargument against land grabs and other practices that were undermining small farmer livelihoods. Its mission is: "The right of every nation to decide on their own farm and food policy ensuring food security in keeping with its traditions and need for sound social and environmental goals." The key elements of food sovereignty include a person's right to food, the priority of local produce and protection against cheap imports, the right of consumers to freely choose what they eat and from whom they purchase their goods, the right of those working in the production of food to appropriate wages, and last but not least, support for sustainable farming.

One thing is increasingly clear: working with nature and not against it, for instance with the terra preta methods, can lead to considerably higher yields and improved farmer income. To achieve this, however, we have to respectfully recognize and learn from the indigenous knowledge of earlier advanced cultures in Europe, Asia, Africa, and the Americas. We, the sower of seeds and consumers of the seed's bounty, must decide how best to feed ourselves, how best to recycle our organic wastes, and how best to maintain

CHAPTER 2

Cultures Need Fertile Ground—The Secret of Black Gold

Advanced cultures need a surplus of food that can only be harvested from fertile soils. This was the case 2,000 years ago in the Amazon region, where huge garden cities flourished on man-made terra preta soils.

I N 1542, THE Spanish conquistador Francisco de Orellana set off on a great adventure. With a group of his compatriots, he navigated along the Amazon and its tributaries in search of the legendary city of El Dorado. During their expedition, they were repeatedly attacked by indigenous tribes. The river was called something else in those days and referred to as the Amazon only later because one of the explorers, the Dominican missionary Gaspar de Carvajal, claimed in his chronicles to have seen female warriors—Amazons—with bows and arrows: "We saw women with our own eyes who fought as warriors in the very front line. These women are very white and tall and have long hair, braided and wound about their heads. They are very robust and go naked, their privy parts covered with bows and arrows." This, as later became apparent, was pure male fantasy.

Francisco de Orellana, in turn, reported millions of people settled on the banks of the river and huge settlements that were separated by "scarcely more than a crossbow shot." Near the mouth of the Tapajós, a major tributary of the Amazon, roughly at the site of today's Santarém, the banks were swarming with people, and the conquistadors were forced to flee. Later expeditions to the Amazon, however, found nothing more than the rain forest. The indigenous population had in the meantime died of pox, flu, and other epidemics the European conquerors had inflicted upon them and against which they had no natural defenses. During these later expeditions, when the riverbanks seemed to be unpopulated and the Amazon women warriors appeared as an invention of an over-imaginative Dominican priest, people began to think Orellana had also been spinning yarns when told of highly populated riverbanks.

THE REDISCOVERY OF TERRA PRETA

FOR A LONG time, anthropologists took it for granted that it was impossible for a highly developed civilization to have evolved in the Amazon rain forests. They argued that the founding of large cities there was inconceivable, as the forest's humus layers were far too infertile to produce food supplies for hundreds of thousands of people. For a long time, this explanation seemed so plausible that no one questioned it.

It is true that most tropical soils in the region contain very little humus. Because of rapid weathering, the ferrous soil is often dyed red, acidic, and very low in nutrients. Organic remains decay quickly, and nutrients are washed away by heavy rainfall.

However, these assumptions were questioned in the 1960s, when researchers discovered conclusive evidence of large pre-Columbian civilizations at the confluence of the Amazon, Rio Negro, and Madeira. Still curious as to how these large populations were able to feed themselves given the traditionally infertile soils, scientists extracted new soil samples in search of an explanation. Eventually, they stumbled across what is now referred to as *terra preta do índio* (Indian black earth).

The black earth soils were generally found on higher ground protected from flooding near the Amazon or its tributaries. Archaeological sites extended from Colombia to Santarém and Manaus to the mouth of the Amazon. Surprisingly, this black topsoil was discovered to be more than 2,000 years old yet was still fertile. The carbon-rich soil, which is now estimated to cover up to 10 percent of the Amazon region, was often 0.5 meter (1.6 feet) thick, and in some places was as deep as 2 meters (6.5 feet).

Brazilian archaeologist Eduardo Neves and other scientists, such as James Petersen, Johannes Lehmann, and Bruno Glaser, found in certain Amazonian soils the remains of charcoal, pottery shards, bones, traces of human feces, ash, and fish bones. Gradually, ancient civilizations were able to create a deep and fertile layer of humus from this wide variety of residual matter.

A tragic incident marred the rediscovery of terra preta. On the afternoon of August 13, 2005, Eduardo Neves and

TROPICAL SOILS

THE TROPICAL RAIN forests of the Amazon basin and Central and East Africa grow, to a great extent, on soils that are referred to as "oxisol" soils. Such soils are heavily weathered and acidic from millions of years of continuous high temperatures and humidity. Because of intensive leaching, they contain very few nutrients or minerals. Another problem with these acidic soils is that the concentration of aluminum is often so high that it is toxic to plants. The agronomic potential of these soils lies in a comparatively thin layer of humus, which releases nutrients and maintains the system in close interaction with forest flora—a perfect cycle. The sensitive ecological system of the rain forests can only grow and prosper, however, as long as the nutrient and water cycles are functioning. When forests are cleared, the thin layer of humus is quickly exhausted and the abundant rains wash away the nutrients, never to return.

his U.S. colleague James Petersen together with two coworkers visited a bar near Iranduba on a peninsula between the Amazon and the Rio Negro for a leisurely beer. Two armed men stormed the bar demanding money, and although the scientists complied, the robbers shot and killed Petersen. Locals were livid with rage and grief. When the criminals were finally arrested after an extensive 21-day manhunt, hundreds of people lined the streets wanting to lynch the murderers. Petersen and Neves were, for the locals, heroes who had discovered pre-Columbian advanced culture, with streets, irrigation systems, and trading routes. Petersen called *terra preta do índio* "a gift from the past." His untimely death robbed him of the chance to further unravel its secrets.

THE CONQUERORS UNDERRATED THE CULTURE OF THE *INDÍGENAS*

THE SECRET TO the large and successful pre-Colombian culture derives from a social system that understood how to live in harmony with nature and use its resources so intelligently that nothing was wasted, much like in nature. The indigenous people valued, in particular, the carbon cycle to create fertile soil. In the process, carbon from charcoal, commonly referred to now as "biochar," had a special role as a storage medium and as a habitat for microorganisms. By using this biochar and recycling all of the everyday waste products, the indigenous population was able to produce extremely fertile soil from the originally poor soils of the tropics and achieve abundant harvests. These stable harvests enabled the founding of cities with population densities that were possibly greater than those in today's Bangladesh, Holland, or Japan.

The European conquerors, however, failed to recognize both the importance of black earth and the cultural achievements of the indigenous population. In the colonization of Brazil, the Portuguese nearly exterminated the indigenous people through forced labor and epidemics, and the new overlords imported black slaves to work their plantations. And even today, many Brazilians look down on their indigenous ancestors, instead of celebrating their successful agricultural traditions. To this day, terra preta research in Brazil is undervalued, though the soils themselves are held in high esteem by farmers lucky enough to have them on their farms.

There are two key differences as to why pre-colonial farming in South America was so different from farming in Europe. First, as we have already mentioned, the humus layer in the rain forests was normally too marginal for farming. Second, in pre-colonial South America, apart from semi-wild lamas and alpacas, there were no large animals for the indigenous people to domesticate. This is, by the way, one of the reasons they were so much more susceptible to flu, pox, and yellow fever viruses than the Europeans, who had become immune to many diseases through living in close proximity to a variety of domesticated animals over the centuries.

South American indigenous cultures, including the Aztecs, Mayas, Anasazis, and Incas, did not practice livestock breeding or use working animals in agricultural labor. This lack of domesticated livestock may have been the reason neither the wheel nor plow was widely used. Probably they didn't even need them, as their garden culture more than catered for their nutritional supplies. Highly developed civilizations foundered with the destruction of this garden culture system by the conquerors, tribal wars, and climatic influences.

INTELLIGENT WASTE MANAGEMENT

SCIENTIFIC STUDIES HAVE confirmed that *terra preta do índio* was man-made. However, up to now, no definitive knowledge or historical records have been found to explain how these anthropogenic soils were created. For this reason, only hypotheses can be formulated and tested experimentally in an attempt to find answers. The following explanations are based on such hypotheses, derived from archaeological evidence as well as insights from historical and contemporary organic land management.

Instead of burning forests to cultivate landscapes, as is the practice in Brazil today, indigenous people of the Amazon cultivated forest gardens among fruit-bearing trees, probably close to their dwellings. While they used the wood from the trees as a raw material for burning or producing charcoal, they knowingly or not, improved the soil over the centuries by mixing food remains, bones, and animal and human excrement with charcoal and wood ash. Gradually, in this way, they increased the soil fertility. The richer the humus, the better the harvest.

An elaborate mixture of crops and intelligent use of waste was necessary to achieve high crop yields over centuries. Ancient forest inhabitants discovered how to recycle the nutrients extracted from the soil via the cultivation of plants so that they weren't washed away by tropical rain and could ensure cultivation and high yields in the long term. All these elements—mixed crops, closing the nutrient cycle, and the formation of humus—are inextricably linked to the emergence of these huge garden cities in the rain forest.

As there was hardly any livestock manure for fertilizing, the human digestive tracts were apparently put to good use for the production of organic fertilizers. Charcoal was probably also an essential tool in preventing diseases and epidemics. Using clay pots as dry toilets and sprinkling charcoal on the contents before making them airtight not only prevents decomposition and unpleasant smells but also reduces flies and the spread of diseases. It is highly unlikely that indigenous people were aware that the black powder so vital to their very survival did not decompose once interred beneath their feet, but this ingenious waste management system proved to be the key to the formation of long-lasting and resilient humus.

People have been producing charcoal for thousands of years, in particular for cooking, which has the benefit of no smoke as compared to cooking with wood. Charcoal is a very good source of energy, and unlike wood, it doesn't rot in tropical conditions. As the indigenous people only had digging sticks and axes at their disposal for tools, and felling a tree with an ax is very demanding, they were more likely to have used timber for construction than for daily cooking needs. Therefore, they probably used scrap wood or dry branches for charcoal production. Charcoal dust is often a byproduct produced while cooking. You can also produce charcoal dust by placing a clay vessel with an opening at the side towards the base in a cooking fire and filling it with firewood, corncobs, or nutshells (see chapter 3, page 92).

Another common ingredient of terra preta is pottery shards from clay vessels of varying sizes that archaeologists have found in all dark, humus-rich soil profiles. The vessels had lids and could be made airtight and were thus suitable for fermenting various different materials. Fermentation, which carries with it a characteristic sweet-sour smell, has been used for thousands of years by many cultures for preserving food—in Europe, for example, in the production of sauerkraut—the preservation of livestock feed, or the preparation of alcoholic beverages.

The original inhabitants of the Amazon basin must have realized the benefits of separating human feces from urine. Diluted urine is a valuable source of nitrogen and potassium but ideally should not be mixed with bacterial feces. The indigenous people possibly collected urine in calabash gourds or clay vessels, using it for fertilizing or other purposes. The special value of "golden juice" was and still is recognized by many communities living in harmony with nature. However, the use of feces can be particularly tricky, as it is easy to spread diseases and transmit parasites. Covering feces with charcoal powder prevents decomposition, regulates moisture, and creates better conditions for lactic acid fermentation. Lactobacilli, excreted by the human intestinal tracts, initiate the fermentation process. The numerous water-filled pores of the pieces of charcoal provide an excellent habitat for these probiotic intestinal bacteria. Harmful bacteria, however, have very little chance of survival, as fermentation has a bactericidal effect and keeps the number of pathogens in check. Parasites are reduced in numbers and roundworms die for want of a suitable host. Through the fermentation process and in the use of airtight vessels, feces can be transformed into a precious raw material that can be stored for long periods of time.

Clay vessels, often beautifully shaped, have been found in terra preta at many excavation sites. The vessels, some containing cooking leftovers, feces, and urine, were as large as 20 liters (5 gallons). Larger vessels would have been heavier and more difficult to transport. Another advantage of having transportable vessels was that family members could have their own toilet, meaning that in the event of illnesses, the risks of infection could be considerably reduced. Whether the original forest dwellers actually had individual toilets has yet to be verified; however, given the high population density, it is likely.

Larger thick-walled clay vessels, sometimes with volumes of 200 to 300 liters (50 to 80 gallons) and often lined in rows, have been found beneath thick layers of black earth. They were possibly fired behind dwellings at a location intended as a garden and probably acted as reservoirs for all the smaller vessels with household wastes. The indigenous people made layers using their charcoal dust mixed with cooking and garden waste, as well as feces and forest soils. Finally, the vessels would have been sealed for the fermentation process and to stop the contents from being washed away or attracting unwanted insects.

In photographs, holes are sometimes detectable in the bases of these excavated clay vessels out of which the fermentation effluent seeped, attracting soil organisms and microbes. They could then enter the vessels through the holes, mix up the organic wastes, and convert them to humus via metabolic processing. The result was precious humus made in a relatively short period that was protected from tropical rain leaching.

Another possible scenario for producing terra preta could have happened as follows: as soon as the vessels began to emit an earthy smell, the indigenous people may have planted a seedling or a banana plant in the vessel to complete the humus-making process. Plant roots live symbiotically with a wide variety of soil organisms that process biological nutrients and supply plants according to their needs. Nutrients cannot be leached, as most of them are bound to the countless soil organisms. In time, plant roots would have burst through the confines of the vessels, providing an explanation for the numerous shards found in deeper soil profiles.

The indigenous people could have stacked the empty vessels on top of the full ones or in rows, filling the empty ones with their daily wastes. They filled the space between the vessels with organic materials, covering them with a coating of earth. Within a short period, they had created elevated beds for forest gardens with soil they could easily till with digging sticks. Soil organisms and plant roots ensured there was continuous loosening and mixing of soil.

This ancient form of waste management effected a steady growth in soil fertility, and unlike today's industrial agriculture, this fertility seemed to increase with the growth in population and the length of stay at a particular location.

CAN TERRA PRETA MAKE A COMEBACK?

REPORTS OF REGENERATION of terra preta soil that had already been dug up are an astonishing phenomenon that has yet to be fully explained. The reports, from Brazil, suggest terra preta can reform beneath the forest within a 20-year period.

How is this possible? How could charcoal be regenerated in the soil? Perhaps the fungus *Aspergillus niger*, which can produce black carbon from dead wood, holds the key. It is a very aggressive fungus that can even attack glass. It seems to be able to transform wood into a carbon structure that can then be colonized by microorganisms. This hypothesis has yet to be scientifically validated, however.

Possibly the original forest inhabitants observed the *Aspergillus niger*, and perhaps they gathered this black fungus from rotting stumps in the forest, propagating it in their clay pots without necessarily fully understanding what they were doing.

The forest garden, with its various levels, can be one of the most productive systems for supplying food. Tall trees provide shade—some also bear edible fruits—and soften the impact of raindrops, particularly during heavy rainstorms. Banana and avocado plants and citrus trees form another group, while on the ground there is room for the so-called "three sisters" crops common among indigenous people in the Americas: corn, beans, and squash. For ground cover, other vegetables and medicinal plants were probably planted. Ground cover vegetation prevents soil erosion and helps regulate water; water evaporates and condenses within a small cycle. The ground, rich in humus, soaks up the water and nutrients like a sponge. Mixed cultivation of the forest gardens supplies energy, building materials, fruits, and vegetables, all within a very small area. Predatory insects have little chance to proliferate, as the plants support and protect each other from all manner of pests.

If we compare terra preta management in the Amazon with our current monocultural agricultural practices, it becomes clear that modern practices cannot compare with the yields per acre of these ancient techniques, at least in terms of total calories produced on terra preta soils. Moreover, current practices promote soil degradation as compared to the regenerative soil practices promoted by so-called primitive cultures. It is clear that twenty-first century horticulture experts and agronomists have much to learn from these long-departed agricultural experts.

THE REDISCOVERY OF TERRA PRETA IN EUROPE

HOW WERE THE secrets of the production of terra preta rediscovered in Europe? It was the fortunate circumstance of interdisciplinary interaction between scientists and practitioners that led to this inspiring rediscovery. Possibly the discovery was already hanging in the air, as apparently many people around the world were trying to unravel the mystery at roughly the same time.

One of these discovery stories took place on the outskirts of Berlin. There, in 2000, Haiko Pieplow and his family built a solar house on an 800-square-meter (8,600-square-foot) plot of land. At that time, he pondered how to avoid wasting energy, water, and food. In addition to installing solar panels, he constructed a wetland system with a filter that would make wastewater drinkable and produce humus. At the same time, various articles and TV broadcasts drew his attention to the phenomenon of terra preta.

In 2005, friends brought him effective microorganisms and he examined their effects on his filtration system (see chapter 3, page 96). The microorganisms improved the purification capacities of gravel filters, as well as boosting the quality and amount of humus produced. In addition, they formed a living bacteria bloom in his active charcoal filter; the charcoal seemed to provide the microorganisms with more than suitable habitation. He was well aware that charcoal had a positive effect on soil fertility from reading old horticultural books and from his experiences as a student of agriculture. For Pieplow, these observations were the key to experiments to reveal the secrets of the origins of terra preta.

In 2005, black earth was made "soil of the year" in Germany. At the same time, some experts were fiercely debating whether black earth had been formed naturally or from deliberate human activities. At the heart of the dispute were the origins of the unusually large amounts of charred carbon that were found both in the black earth of certain steppe regions and also in terra preta.

Haiko Pieplow discussed his observations from his experiments with a network of scientists and practitioners whose expertise ranged from permaculture to organic farming to sustainable community water management. Terra preta researchers Bruno Glaser, Johannes Lehmann, and William I. Woods; landscape ecologist Willy Ripl; and humus researcher Herwig Pommeresche provided key guidance.

Glaser, Lehmann, and Woods had already compiled and published many findings about Amazonian terra preta, providing a baseline for the physical and chemical processes of what would soon come to be called biochar. Their observations on soil biology were particularly valuable. Ripl drew

THE BLACK EARTH OF THE STEPPES AND BÖRDE

THE RUSSIAN TERM "chernozem" describes black earth soils found in the fertile Eurasian steppes of Ukraine and southern Russia, in the lowlands of Germany, and in some parts of the North American prairies.

Chernozem evolves on rich chalky, granular soils, such as loess under the climatic conditions of the steppes. Regular prairie fires, which at least in the North American prairies were deliberately started by the local populations, created large amounts of charcoal. Grasses provide the source material for humus. Because of periodic cold and dry spells of the steppes, these grasses decompose very slowly and accumulate over time. Humus and charcoal from the fires turn the topsoil black. The burrowing activities of soil fauna, which tunnel deep into the soil in the cold winters and hot summers, transport the carbonized matter and humus-rich layer deep below the soil surface. All of this makes the soil extremely fertile.

However, terra preta and some of the other rare dark-colored soils, such as plaggen soil, differ from chernozem because they are anthropogenic (i.e., man-made) and not created only through natural processes. In this book, when we talk about black earth or terra preta soils, we mean those which were created with the help of ancient civilizations and are beginning to be created once again by those focused on regenerative agriculture and gardening.

attention to the importance of water and trees on soil formation, and the development of landscapes and climate. Pommeresche provided information about the high productivity of forgotten advanced cultures and the importance of humus for plant nutrition, prompting experiments in his own garden.

Joachim Böttcher, who had built Haiko Pieplow's filtration system, was enthusiastic about the results of the first plant pot trials using terra preta subsoils. Together with Alfons Krieger, an experienced organic farming consultant, he refined the experiment, and in 2010, he built facilities for producing terra preta subsoils at a farm in Rhineland-Palatinate.

While all this was happening, knowledge about the production of terra preta in Europe was progressing rapidly. Farmers and gardeners with experience of lactic acid forming effective microorganisms began experimenting with the production of terra preta. Farming consultant Christoph Fischer started a regional black earth project with farmers in Upper Bavaria (see chapters 3 and 4). In Berlin and Brandenburg, Jürgen Reckin (agricultural researcher) and Marko Heckel (agricultural consultant) were also experimenting with the production of black earth.

In the last 10 years, many regional terra preta interest groups have sprung up all over the world and plenty of research projects came into being. Established agricultural science, however, has sometimes seemed hesitant to pick up on the opportunities of long-term storage of carbon by encouraging the formation of humus, possibly because of its dependence on the agro-industry.

Bruno Glaser, from the Department of Soil Biogeochemistry of Martin Luther University of Halle-Wittenberg, and Claudia Kammann, from the Institute of Plant Ecology at Giessen University, are two distinguished exceptions and are among the leading scientists involved in international biochar research. Glaser was also the first to prove, in the early 2000s, that Brazilian terra preta was man-made and contains significant amounts of human waste.

Nowadays, information and discussion forums about terra preta are flourishing on the Internet, and there is an increasing number of biochar conferences being held around the world. There is still a long way to go before the full potential of biochar is realized, but the growing exchange of insights and experiences is critical to increased adoption of the practice of making high-fertility soils. Recent attempts by individuals to patent partial understandings of the production of terra preta seem self-serving, counterproductive, and disrespectful to those who actually invented the process for making these fertile soils. Terra preta is not the discovery of a single person but an ancient cultural heritage of humankind that should be treated in a dignified manner. As a forgotten and now rediscovered cultural technique, it should be there for the good of all. In the meantime, thanks to the Internet-based exchange of knowledge, it has become almost impossible to "enclose" this "common land" with patents. This exchange makes it possible to share knowledge of sustainable farming practices worldwide.

Some scientists believe, based on indicators from Central Africa, that knowledge about the production of black earth was passed on from mother to daughter. Such knowledge, unfortunately, was rarely passed down in writing. For thousands of years, women all over the world have developed and been responsible for tending gardens. They were and are, in many places still today, responsible for cooking, food, and hygiene. Based on their historical gender role, women often think more than men about feeding the family and thus the future of soil that produces their food.

THE DECLINE OF CULTURES THROUGH EROSION

IN HIS BOOK *Dirt: The Erosion of Civilizations,* the U.S. geologist David R. Montgomery points out the close links between the decline of cultures and the erosion of soil. A surplus of food that can sustain a "thinking" class of administrators, writers, engineers, and philosophers is a prerequisite for the emergence of advanced cultures, and such surpluses can only be achieved on fertile soils. If soil fertility isn't steadily cared for and maintained, civilizations will inevitably decline.

"In a broad sense, the history of many civilizations follows a common storyline," writes Montgomery. He explains that most cultures emerge from fertile valleys, where populations grow gradually. After continued growth, inhabitants are forced to cultivate the surrounding hills and mountains, where the arable land, created by the felling of trees, is quickly borne off by rain and wind.

Sometimes soil degradation is caused by other problems. In Mesopotamia, the cradle of human civilization and cultivation, increasingly saline soils were caused by groundwater containing too much salt. By watering fields selectively or allowing them to lie fallow for a while, problems with salt

THE FLOATING GARDENS OF MEXICO CITY

THE FLOATING GARDENS of Xochmilco on the southern outskirts of Mexico City were once considered the eighth wonder of the world. In 1987, UNESCO granted the remains world heritage status. There too the descendants of the Aztec farmers work with biochar. The Nahuatl word for biochar, "xochmilco," means "where the flowers grow." The nowadays greatly reduced gardens are found between numerous branched canals providing water to the fertile areas, where flowers, vegetables, and fruits are planted.

This lush garden culture at one time provided Tenochtitlan, the densely populated Aztec capital, with sustenance, before the Spanish conquistadors almost completely destroyed it and proceeded to build Mexico City on its ruins. Before the arrival of the Spaniards, the floating gardens were spread across a huge lake. The Aztecs covered reed rafts with mud from the bottom of the lake, using animal and human excrement as fertilizer and anchoring them with willow roots. In time, they took root in the lake bed and have now become islands.

This special earth/humus mixture, similar to terra preta, was the basis of perhaps the world's most fertile water-garden culture, producing several harvests per year. Approximately 30 square meters (320 square feet) are enough to feed one person for an entire year. By comparison, today's industrial agriculture requires nearly 1,800 square meters (19,000 square feet) in tropical climates or 2,800 square meters (30,000 square feet) in colder climates to achieve the same returns.

Although the Spanish conquerors almost completely destroyed the lakes and canals, the "chinampas," as the floating gardens are called, still harvested some 500 tons of fruit, vegetables, and flowers and delivered them to the markets of Mexico City by boat until the 1950s. Sadly, today the waters are contaminated by toxins, the past diversity of cultivated vegetables has completely disappeared, and the cultivated areas have severely declined. This once-bountiful world heritage site is now under threat.

Lettuce, fennel, spinach, and radishes continue to grow on the remaining areas. A commercial representative of the 1,600 farmers who work the land there told a group of visiting farmers from Chiemgau: "The fields are magical." Their Aztec ancestors had scooped up mud from the lake bed for centuries and managed to be extremely productive without using any fertilizers. When necessary, the fertility of the soil was improved by using charcoal powder.

could be alleviated. However, in the land between the Tigris and Euphrates, centuries of high productivity of the fields enabled the population to increase so drastically that the intensity of cultivation and thus irrigation had to be increased. This led eventually to the contamination of the soils by increasing levels of salt.

In this respect, the ancient Egyptians were better off. The regular flooding of the Nile meant that the fields were annually coated with fresh fertile mud, allowing the salt to be washed out to sea. Because of this natural renewal of the soil, the culture of the pharaohs could be maintained over many centuries, without ever losing the fertility of their fields. Not until the nineteenth century and the beginning of the cultivation of cotton for export to Europe did the natural rhythm of farming become unbalanced, and this was simply because cotton needed irrigation throughout the year, causing increasing salt levels. The construction of the Aswan Dam in the 1960s meant that the fertile soil carried down the river from the highlands of Ethiopia is now deposited on the bed of Lake Nasser. To offset the loss in naturally replenished nutrients, Egyptian farmers now apply an inordinate amount of synthetic fertilizers to their fields. As a result, the descendants of this proud ancient agrarian culture can no longer feed themselves and must now rely on imported food to survive.

One of the reasons the city states of ancient Greece and the Roman Empire declined was because their inhabitants mercilessly exploited the hillsides and slopes in their quest for arable farmland and timber. Around 590 BC, the population of Athens could no longer sustain itself because the surrounding hills had eroded, and by 400 BC, many Greek cities were reliant, to a considerable extent, on imported cereal crops from Egypt and Sicily. In imperial Rome, too, contemporary chroniclers noted a higher rate of erosion. At the very beginning of the first millennium, Rome had to ship in an estimated 200,000 tons of cereals from its colonies. Erosion continued to forge on, eventually affecting even the Roman provinces.

The indigenous people of what is now North America were not immune to soil problems. The Mayas, who lived in the area of today's Yucatán Peninsula in Mexico, also contributed to the collapse of their once-great culture. According to recent studies, the Mayas' high-production farming fed up to 20 million people, but they eventually fell victim to deforestation. Ever-increasing networks of irrigation canals could no longer prevent the land from drying up. At around 900 AD, at the peak of their civilization, a 30-year drought began that probably led to a dramatic decline of their population, of which a mere 10 percent survived. After this enormous population decline, the jungle returned and the former city centers became overgrown.

Modern Europe, too, can only feed its population by relying on imported food and animal fodder. Since the 1800s, European powers have exploited their overseas colonies, importing mostly luxury goods, such as coffee, tea, and sugar. Today, however, the landscape is very different. Europe's land area would have to increase by one third to enable self-sufficiency when it comes to the production of food, animal feed, oils, and biofuels.

BLACK EARTH IN OTHER CULTURES

THERE ARE NUMEROUS indications that other cultures had discovered the secrets of creating humus with charcoal. Language barriers have until recently kept hidden the fact that traditional farming practices in China, Japan, India, and Korea all employed the benefits of using charcoal in agriculture.

In Japan, for example, charcoal has been made for centuries from rice residues. Rice husks are carbonized and the resulting powder worked into the ground. A related process was called *haigoe*, which Japanese farmers used as a fertilizer consisting of human waste, including feces, and charcoal powder.

The Japanese, additionally, had developed many ways of preserving food—such as rice, beans, and fish—by fermentation. In all probability, in the past they also fermented cooking wastes in clay pots to produce high-grade fertilizers.

The word "bokashi," known today by an increasing number of gardeners, also comes from the Japanese and means something like potpourri. Bokashi involves converting organic matter by fermentation into a microbial more stable form with an increased nutritional value. This is achieved through a process by which many precious vitamins, enzymes, amino acids, and antioxidants (free-radical scavengers) are created. In a similar way to humans who find it easier to digest sauerkraut than raw cabbage, soil organisms can better digest and metabolize fermented organic wastes.

Other fermentation techniques are just as old as the bokashi method. In the tropics, where hurricanes and tornados can quickly eradicate harvests, people have for thousands of years fermented breadfruit in leaf-lined pits to help them survive such catastrophes. When these pits weren't needed anymore, sumptuous plants started to grow in them— something people who were used to observing nature could not have missed.

Similar evidence is provided by the luscious plants growing in the vicinity of charcoal and earth ovens. Many indigenous societies still use earth ovens for cooking, and the remains of charcoal scattered around the mouth of the ovens enriches the soil.

Terra preta is not something that can be achieved in a single season, with a single application. True terra preta soils are formed only in places where soil life is appreciated, encouraged, and maintained. Those indigenous cultures that venerate Mother Earth as a productive force figured this out and were able to nourish their growing populations for as long as they nourished their soils, barring unforeseen invasions! Cultures that fail to learn how to practice sustainable farming, noted Montgomery, sooner or later, founder.

PART II

THE HANDBOOK

CHAPTER 3

Climate Gardening— The Basic Principles and Materials

Through the use of climate gardening, huge amounts of carbon can be permanently stored in the fertile humus layer. Biochar and the stimulation of an active soil life are indispensable.

WITH RAPIDLY ADVANCING climate change in the twenty-first century, extreme weather has become the new normal, threatening whole regions with drought, wildfires, flooding, increased pests, and more. Wildfires rampaged through Russia in 2010 and the western U.S. in 2015; massive floods devastated Pakistan in 2011 and China, India, Nigeria, and North Korea in 2012. Droughts around the globe continue to plague farmers, especially in the western U.S., Australia, India and Africa, leading to serious depletion of water for short-term growing needs.

Organic gardeners and farmers go about their business in a considerably more climate-friendly way than conventional farmers, but they still emit greenhouse gases. If farmers were to implement large-scale humus regeneration projects by means of terra preta techniques, global warming and its consequences could be reduced and food security could be improved in many parts of the world.

LONG-TERM CARBON STORAGE

HOW CAN THIS be achieved? Every plant absorbs solar energy and carbon dioxide, using them to develop biomass. When plants die, in most cases all of the assimilated carbon is released back into the atmosphere within a short period of time as part of the normal carbon cycle. But when these organic residues are carbonized via pyrolysis (i.e., the application of heat in the absence of oxygen), 25 to 50 percent of the carbon is converted into long-lasting biochar carbon, which is highly resistant to degradation for centuries. When this long-lasting biochar carbon is worked into the soil as a soil amendment, it can enable highly fertile humus to develop. This gives us a win-win-win situation for the climate, soil life, and food security.

Biochar is made from biomass using climate-neutral pyrolysis. The heat needed to sustain the process is supplied by energy-rich syngas, a byproduct of pyrolysis. Depending on the original biomass and production parameters used to make biochar, biochar may contain up to 95 percent pure carbon. As this carbon is resistant to microbial degradation, it sequesters carbon when applied to soil and thus helps to decrease the level of carbon in the atmosphere.

Globally, biomass absorbs approximately 120 gigatons of carbon from the atmosphere each year through photosynthesis. A similar amount is returned to the atmosphere through biological respiration, decomposition, and natural fires. However, on top of this, humans release nearly 9 gigatons of carbon in the form of carbon dioxide every year through burning fossil fuels, such as oil, gas, and coal. These extra 9 gigatons from fossil sources is what is causing the greenhouse effect leading to climate chaos. Long-term stabilization of carbon via pyrolysis, combined with an increase in global biomass production through reforestation and revegetation, is an ideal way to draw down a significant amount of CO_2 from the atmosphere. Modern pyrolysis facilities can char organic residues so efficiently that the equivalent of 500 kilograms (1,100 pounds) of carbon dioxide can be sequestered for every ton of woody biomass. Amending a soil with 1 ton of biochar that contains an average of 65 percent carbon keeps at least 2 tons of carbon dioxide out of the atmosphere.

Greenhouse gases can be further reduced by the fermentation of organic wastes, as in the making of terra preta. Lactic acid fermentation prevents the rotting of biological residues. Without rotting there is no stench and no release of methane or nitrous oxide.

THE CLIMATE CATASTROPHE CAN BE HALTED

EVERYONE IN THE world is contributing to global warming to some extent. However, the carbon footprint varies enormously from one country to the next, as well as from one person to the next. Some consume excessive amounts of oil and gas; others use mostly renewable energy. Some jet around the world, whereas others opt for using a bicycle.

HUMUS AND THE CARBON CYCLE

AT PRESENT, CARBON levels in the atmosphere are increasing at roughly nine gigatons a year. The primary causes for this is the burning of fossil fuels, slash-and-burn forest clearing, and agriculture, which is responsible for various greenhouse gas emissions. Plowing, tilling, and fertilizing with chemical nutrients will, in the long term, damage soils, disrupt soil organisms, mineralize organic matter through contact with the air, and release CO_2 and other greenhouse gases. Whenever humus or peat is lost, greenhouse gases are released. In contrast, whenever humus is created, carbon is sequestered and thus there is less CO_2 in the atmosphere.

Some consume vast quantities of meat, whereas others try to buy locally produced organic food or follow a vegetarian diet.

One ton of carbon corresponds to 3.6 tons of carbon dioxide. A 100-square-meter (1,000-square-foot) plot of land with a humus content of 1 percent stores approximately 0.3 tons of carbon, or roughly 1 ton of carbon dioxide. Therefore, for every additional percent of humus stored in the same area, one additional ton of CO_2 can be sequestered; humus content of 10 percent thus equals 10 tons. If a gardener succeeds in increasing the humus content in the garden from 2 percent to 10 percent, then the equivalent of 8 tons of carbon dioxide can be stored on the 100 square meters (1,000 square feet). An increase of 20 percent would sequester an impressive 18 tons, slightly more than the annual carbon footprint for an average American.

A few examples will show how important the making of terra preta soils could be in the effort to rebalance atmospheric carbon. In Germany, there are more than 1 million urban garden plots, spanning more than 46,000 hectares (113,000 acres). If all these gardeners were to increase the humus content of their plots from 2 percent to 10 percent, the equivalent of 30 million tons of CO_2 could be sequestered or drawn out of the atmosphere and safely stored under our feet. All these sequestration efforts come with the added benefit of improving soil productivity, thereby enabling more food and flowers to be grown.

In total, German farmers cultivate roughly 12 million hectares (30 million acres) of land. If they all began to recycle various agricultural and organic urban wastes and added the carbonized and composted remains to their soils, thus increasing the humus content of their soils by 4 percent over the next 20 years, up to 5 billion tons of CO_2 equivalent could be stored permanently in the soil—that is more than 5 times the total annual CO_2 emissions of Germany. Of course, we understand this calculation is somewhat unrealistic, since not all farmers would participate, nor is there currently sufficient biochar production technology or excess biomass to meet these goals, but more and more production technologies are becoming available that can efficiently process a wider range of biomass, including sewage, manure, digestate, and more. Although the increase of organic matter in agricultural soils cannot save the climate without massive reduction of greenhouse gas emissions, it can substantially help to recalibrate the global carbon balance during a transition time.

The standard small to tiny garden plots are not sufficient to compensate for the average CO_2 emissions per person per year. You can, however, improve your own personal CO_2 balance with climate gardening or in a number of other ways. First, by the already mentioned storing of carbon in the soil using terra preta techniques, where an increase in humus content of 10 percent is feasible within a couple of years. Second, by organically producing your own vegetables and fruit, which also reduces CO_2 emissions through reduction of chemical fertilizers, packaging, transport, and costly storage. Third, by avoiding or reducing waste disposal and using your own organic wastes for compost or humus production. In a garden, there is no such thing as too much humus!

Leberecht Migge, urban garden visionary, was an important early advocate of returning all organic waste to the soil.

In 1919, he proposed that all city dwellers should have the option of self-sufficiency. He believed a family of five could, more or less, cater for their needs on a 400-square-meter (4,300-square-foot) garden plot. He figured out that this amount of land could yield more than 1,200 kilograms (2,600 pounds) of vegetables and potatoes and 300 kilograms (660 pounds) of fruit, as long as the family plowed all organic wastes back into the ground. He emphasized that fertile garden soils could be easily produced and recommended using human feces. Migge, too, invoked the tried and tested experiences of ancient times and various cultures.

In California, a family of four that lives on a mere 4,000-square-foot plot in an urban setting is in fact harvesting 6,000 pounds of food from their urban farm, including vegetables, fruit, eggs, and meat. The Path to Freedom experiment, which started more than 10 years ago, not only nourished this family but also inspired millions of other urbanites to convert their sterile lawns into fertile mini-farms.

If the humus content were increased by 0.1 percent per hectare per year of agricultural land, the equivalent of 10 tons of CO_2 per hectare could be absorbed by the soil. Such humus growth rates could continue for 15 to 20, or perhaps 30 years. Canada could compensate more than half of its 730 million tons of annual CO_2 emissions with an annual increase of 0.1 percent of humus on its 43 million hectares of arable land. If done on a global scale, the 1.4 billion hectares of the world's arable land could sequester annually the equivalent of 14 billion tons of CO_2, corresponding to 40 percent of global greenhouse gas emissions.

Urban climate gardening benefits the climate in another more nuanced way. In summer, luscious gardens, wall-climbing plants, and flowers in window boxes ensure a cooler atmosphere, because evaporating water is a very effective coolant. Rainwater and groundwater are purer, because urban gardens enhanced with terra preta can filter, absorb, and render toxins unavailable to plants. Dust concentrations, particularly fine dust pollution from traffic, are reduced, because urban dust tends to settle on leaves but becomes less harmful thanks to the higher humidity of the air. Biodiversity also increases (see chapter 5).

Community gardening also improves the social climate between individual gardeners. Often, the most important benefit of climate gardening is the joy that it brings each individual. For many, gardening with enhanced terra preta soils is sensuous, ethical, fruitful, healthy, and can lift the barriers between people and nature, while helping to heal the carbon imbalance. At its very essence, climate farming cultivates hope and happiness, something that is sorely needed in this era of climate chaos.

HUMUS PRODUCTION BY CLIMATE FARMING

HUMUS NATURALLY DEVELOPS from decaying plant residues and animal remains, as well as the metabolic products of plant roots and soil organisms. The rate of production and specific content of humus varies greatly depending on soil type, location, climate, and cultivation practices. Forests in Central Europe have comparatively high humus contents of 10 percent. If the trees were felled and the areas then used

CLIMATE FARMING IN VALAIS

SWARMS OF BUTTERFLIES—RED admirals, fritillaries, and other even rarer species—flutter among the flowers seeded amid grapevines. The Mythopia vineyard, a contraction of "myth" and "utopia," lies among the majestic 4,000-meter (13,000-foot) Alpine peaks of Valais, Switzerland.

Since 2005, under the management of Romaine Bonvin, bio-diversity has increased rapidly at Mythopia. To create a small paradise for beneficial insects, bacteria, and fungi, an assortment of fruit trees—apple, quince, peach, cherry, apricot, almond, and fig—were planted in the vineyard at 50-meter (160-foot) intervals. Boxes with holes bored in them offer "hotel" accommo-dation for wild bees and wasps—useful creatures that feed on insects that would otherwise harm the grapes. You can smell the fragrance of thyme and lavender. Between the vines, wild herbs, roses and marigolds, ancient cereal species, tomatoes, squash, and other vegetables grow. This unusual symbiosis is beneficial also to the vines, as it increases resistance to pests and increases wine quality.

High biodiversity is just one distinctive feature of Mythopia; another is that climate farming is practiced here. Since 2007, the first large-scale field research into biochar in Europe has been taking place on an area covering 3,000 square meters (9,800 square feet). Biochar used at Mythopia was supplied by Europe's first pyrolysis unit specializing in biochar production, run by Swiss Biochar and a number of partners near Lausanne. The biochar was then biologically "activated" through composting with animal manures and vine residues. Following an intensive composting regime of 6 weeks, black earth soil substrates are ready to be taken out and applied to the vineyard roots.

Mythopia has coordinated large-scale field trials with biochar since 2008, not only in vineyards but also with a growing number of different crops as well as agroforestry systems. They compared the effect of various blends of biochar with organic nutrients, such as composts, animal manure, urine, and green manures. "Biochar without organic nutrients is like a sheet of paper without words," as they say in Mythopia. The biochar organic blends and macerations produce both qualitatively and quantitatively improved results for a variety of crops and provide ready-made biochar products for several companies in Switzerland and Germany.

High up on the upper limits of the vineyards, there is a garden with a glorious panorama of the surrounding mountain landscape. It was set up by the head of an anti-addiction program; here, people with former drug dependency problems cultivate zucchinis, leeks, and lettuce, all on the finest, healthiest black earth—their horticultural success contributes to their recovery. Already, in its first year of cultivation, the 400-square-meter (4,300-square-foot) garden yielded vegetables and fruits for the community kitchen with a value of more than 2,000 euros.

for pastures, humus content would sink to 5 or 6 percent. If the areas were used for arable farming, the content would decrease even more, to about 2 to 3 percent.

The loss of humus in agricultural soils is not just a modern phenomenon. In the early days of farming, more than 10,000 years ago, billions of tons of carbon were lost worldwide through deforestation, erosion, overgrazing, and salination. Since the invention of tractors and synthetic fertilizers, humus losses have accelerated at an alarming rate. A major portion of lost humus leaks into the atmosphere in the form of CO_2, thus having a considerable negative impact on climate change.

As good as organic farming is for the environment and for consumers of organically produced food, it is limited in terms of how much it can raise and maintain the humus content of soils. Depending on clay content, healthy organically farmed soils generally contain between 3.5 and 6 percent humus. It should be noted that this level of humus can only be retained through the use of sound agricultural practices: closed-loop cycles of organic matter and fertilizers, undersowing of secondary crops (e.g., seeding legumes under corn), cover crops, organic fertilizers, and no-till farming (i.e., a way of growing crops without disturbing the soil through tillage).

The incredible potential of terra preta methods lies in the fact that the humus content is 15 percent—or in some cases even higher! Many historical and current examples all over the world demonstrate that it is possible to achieve these levels by carbonizing organic waste material and incorporating it into the soil with organic nutrients such as compost or manure. Although some argue that there is insufficient biomass available to make a substantial impact on climate change, the reality is that there is more than enough organic waste available that currently goes unused. In many cases, the disposal of such non-recycled organic waste is a burden to the environment and contributes to climate change by way of increased transportation costs and greenhouse gas emissions from landfills and other sources.

The reality is, we are massive wasters of biomass; we waste up to 40 percent of all food produced, we waste lawn cuttings by tossing them in garbage cans instead of recycling these nutrients into the soil, and we waste fall's bountiful foliage when we dump it in with our household wastes. All of this organic waste can be carbonized or composted and combined to make humus, which can regenerate our failing soils and rebalance atmospheric carbon levels.

Chemically farmed monoculture fields experience especially rapid humus loss. The most intensively farmed soils in temperate zones of Central Europe, Asia, and North America now have humus levels of 1 to 2 percent for this reason; some regions, such as the sandy soils of northern Germany, Spain, and Portugal now have almost desert-like conditions, with less than 1 percent humus. The critical threshold for humus content, below which harvest volumes significantly decrease, is 2 percent for temperate climate zones and 1 percent in the tropics. At levels below this, the capacity of soil to retain water declines drastically, meaning that plants have difficulty surviving longer dry periods. Although temperate areas, with humus values of 1 to 2 percent, are able to produce reasonably stable yields through the use of chemical fertilizers and

MEASURES FOR ENCOURAGING
HUMUS FORMATION THROUGH CLIMATE FARMING

FERTILIZING WITH COMPOST (rather than commercial fertilizers or slurry)—compost is ready-made humus and thus helps in the regeneration of humus.

Using biochar fertilizer to retain organic nutrients—especially in combination with liquid organic nutrients, such as animal and human urine, liquid digestate, and manure.

Minimum tillage (rather than using plows, cultivators, and harrows)—the less the soil is disturbed, the lower the oxygen transfer and the more stable the existing humus.

Permanent cover of the soil, either through mulching or through cover plants (rather than soil preparation in fall or winter fallows)—only green covering of the soil can provide nourishment for soil organisms, especially in winter. The use of leguminous plants is particularly good for this purpose, as they assimilate atmospheric nitrogen and help establish humus.

Crop rotation (rather than monocultures)—by increasing plant diversity, the root varieties and the stability of microorganisms are also boosted. This creates a precondition in the soil from which humus can develop.

Intercropping (rather than monocultures)—the simultaneous planting of multiple crops, which can grow at the same time and positively influence each other. The increase in root diversity stimulates humus formation.

Agroforestry—combining tree crops and stable crops where leaf fall, root exudates and decaying root parts stimulate humification, soil life, mineral and water cycling.

Avoiding measures that lead to humus degradation—these are above all: the use of commercial fertilizers, soil tillage, monocultures, and the use of pesticides.

Support soil biology using the previously mentioned measures—humus production is a vital process that needs to be maintained.

irrigation, this is no longer possible in the heavily depleted soils in the tropics, where yields are declining every year, resulting in diminished financial returns from chemical farming. The soil becomes exhausted; small farming families can no longer provide for their needs; and malnutrition and starvation set in. This is the case in many African countries, in Southeast Asia, and increasingly in Latin America. Nearly 90 percent of the starving billions live in regions where the humus content of the soil has sunk below critical levels.

Rattan Lal, from the Carbon Management and Sequestration Center in Columbus, Ohio, is one of the most commendable researchers in the field of carbon management. His field studies over many years in Africa, Middle America, and Southeast Asia show that the harvest volumes increase on a linear basis in accordance with increases in humus content. This applies in particular to soils with the critical humus levels of 0.5 to 2.5 percent. His conclusion: an increase in humus content using sustainable farming methods can secure food supplies for a growing world population. Every additional ton of carbon in a hectare of arable land in the tropics and subtropics can significantly increase the yields per hectare—between 20 and 70 kilograms (44 to 150 pounds) for wheat, 10 to 50 kilograms (22 to 110 pounds) for rice, 30 to 300 kilograms (66 to 660 pounds) for corn, and 40 to 60 kilograms (88 to 130 pounds) for beans.

Climate farming can without a doubt achieve annual humus increases in arable soils of 0.045 percent, which corresponds to the above mentioned 1 ton carbon per hectare. The humus level could even increase by 3 to 5 times this value if the climate farming methods listed in the sidebar were rigorously promoted globally. This would correspond to sequestering between 10 to 20 tons of CO_2 per hectare per year, over a period of at least 20 to 30 years.

How might this happen? One straightforward way would be to scrap current subsidies for the use of chemical fertilizers and implement incentives, as well as providing investment capital for humus formation. In addition, establishing organic agricultural colleges and consulting networks, and making knowledge about the production of substrates with biochar accessible to all would greatly contribute to a thriving humus production culture. Raising the humus content of soil using natural methods isn't complicated or mysterious; the know-how has existed for millennia—what is missing is the will to realize it. Sequestering carbon through farming is not only the most expedient and the most ecological way to reduce CO_2 levels in the atmosphere; it is also the most cost efficient.

Since humus formation leads to a reduction in CO_2 levels in the atmosphere, if we could increase global soil carbon levels by 10 percent in the next 100 years, we could safely sequester the equivalent of 800 gigatons of CO_2 beneath our feet. This effort would boost food security and at the same time help decrease CO_2 levels in the atmosphere back to safer levels. Provided that in parallel to this effort, other steps were taken to remodel our current society into a more carbon-balanced and ecologically oriented economy, it is feasible that the level could revert to preindustrial status.

Johannes Lehmann, from the Soil and Crop Sciences Department of Cornell University, talks of "a black revolution." If one-third of global harvest wastes were transformed into biochar, greenhouse gases could be reduced by 10 to 20 percent. Other researchers quote even higher figures. Australian ecologist Tim Flannery wrote in an open letter on biochar that "biochar may represent the single most important initiative for humanity's environmental future," adding that it is "the most potent engine of atmospheric cleansing that we possess."

HOW TO MAKE BIOCHAR

FROM THE TIMES of the Bronze Age, humans have been producing charcoal and using it either for low-smoke cooking or for producing and tempering metals. Biochar is made in a similar fashion to charcoal, though generally at higher temperatures, which allow more volatile materials to burn off but result in slightly lower yields. Although the two terms are often used interchangeably, the true difference is in the end use. Charcoal is combusted for cooking or other uses of the resulting heat, whereas biochar is the term used when the material is used in a way that its carbon is preserved as a soil amendment, animal feed additive, or building material, in textiles or in batteries.

Biochar has a much higher energy density than the biomass from which it is made. It does not rot and, unlike wood and other biomass, can be stored for long periods in warm, moist conditions. A traditional form of biochar production is the carbonization of wood in charcoal kilns. Charcoal burners pile up logs and thick branches, cover them with earth, and let them slowly smolder without oxygen. Those ancient charcoal kilns are generally not built to take advantage of the massive amounts of heat produced for other purposes, have very poor air quality controls, and are certainly not recommended to be used for climate farming. And if terra preta soils all over the world contain millions of tons of biochar, it is impossible that it was produced only from wood, which is so difficult to turn into basic tools. Our ancestors must have had other methods to produce biochar.

In 2014, Hans-Peter Schmidt and Paul Taylor wanted to show how our ancestors were able to produce, with simple means and without high technology, large quantities of biochar. Additionally, they sought a simple, inexpensive, easily adaptable technology for terra preta projects in the developing world. If earlier peoples in South America, Australia, Scandinavia, Palestine, China—actually almost everywhere—were able to produce and apply such quantities of biochar that their soils were partially blackened throughout, this must be achievable today in even the poorest tropical countries. They further hoped to develop a technology that would allow farmers and gardeners in rich countries to convert their own residues into biochar as an alternative to buying it.

They began experimenting with the ancient technique of smokeless fires and combined this with the observations of archaeologists, namely that black soil deposits are often found in soil profiles as clearly demarcated cone pits with an upper diameter of about 2 meters (6.5 feet) and a depth of 1.5 meters (5 feet). At first they suspected these soil cones were simply rubbish pits, which when filled, were burned from the top down, only to be replenished again, but they started to think what if these human-sized pits were used

are comparatively small and most suitable for gardeners and hobbyists, but the principle is clear: produce biochar using the fire and not by suppressing it. Schmidt and Taylor also took as inspiration the form of fire containers that were used throughout the East for religious sacrifices. Under the name of Agnihotra, the Vedic fire ritual, they are still widely used today in India. The size of the Agnihotra bowls is generally small, but for temple rituals, there were larger fire bowls made of copper. The dynamics of smokeless flames over the fire pits, dancing to the heavens, clearly showed that Schmidt and Taylor were on the right track with the physics of fire.

Based on these principles, which may mark a U-turn of the modern direction of pyrolysis for farm-scale biochar production, they were close to developing an optimized low-cost biochar kiln for the production of high-quality biochar in large quantities and at very low cost. The first principle of biochar craft is this: use the pyrolysis gases as cover gas and thus create with the fire the air exclusion for pyrolysis.

This fundamental principle was the starting point for their design of the Kon-Tiki, an open conical kiln for making biochar. They chose the name Kon-Tiki in memory of Thor Heyerdahl, who asserted in the 1940s that the ancient inhabitants of South America were able to cross the Pacific to Polynesia in handmade boats. Experts virulently attacked Heyerdahl's theory, until he silenced them by building such a boat with only the tools and materials of the South American natives and crossing half the Pacific from Lima to Polynesia. He named his boat Kon-Tiki after the South American god of sun and fire. Schmidt and Taylor's goal was quite similar, though not nearly as adventurous. They wanted to show that only craft and no monopoly-owned high tech is necessary to create highly fertile humus-rich soils with biochar.

KON-TIKI CONE KILN

ALTHOUGH THE BIOCHAR quality from the first experiments with an excavated earth kiln looked pretty good, it was too inhomogeneous for standardized products. The open combustion of the pyrolysis gases was fairly clean but not always stable, especially in gusts of wind, and Schmidt and Taylor were not able to completely prevent the emergence of smoke. They had to go one step further to study the operating principles more precisely and optimize the different parameters of the system. At this stage, they designed and built the first 750-liter (200-gallon) aboveground Kon-Tiki made out of steel.

With an upper diameter of 1.5 meters (5 feet), a height of 0.9 meter (3 feet), and a wall inclination of 63 degrees, a steep cone shape was chosen so that the resulting biochar was well compacted and would make a consistent fire front at the surface for a reliable barrier to oxygen. Unlike the earthen walls in the earth kiln, the steel walls reflect the pyrolysis and combustion heat back into the kiln, resulting in a more uniform temperature distribution and thus ensuring more homogeneous charring conditions and resulting biochar quality. More importantly, the decisive criterion for the success of the new steel shape was the difference in combustion dynamics with the change from a sunken to an aboveground form. They found that the combustion air that is drawn down onto the burning surface is preheated as it rises along the hot outer wall of the kiln. Preheating the combustion air significantly reduces the cooling of the unburned gases, generating more stable combustion dynamics and greatly reducing smoke production.

Once the kiln reaches its working temperature of 650 to 700 degrees Celsius (1,202 to 1292 degrees Fahrenheit), hardly any smoke is visible. The combustion air rolls in over the metal edge of the outer wall and into the kiln. But at the same time, the burning gases must escape upwards, so, similar to a clockwork, a counter-rotating vortex is established in the center of the kiln. Thanks to the establishment of this horizontal vortex, the air supply to the fire zone is stabilized. The wood gas, which is heavier than air, is kept in the vortex until it is completely burned. Thus, the second fundamental principle of the Kon-Tiki kiln is the development of a horizontal gas-air vortex, which provides a stable, smokeless combustion regime.

As we had observed with the open earth kiln, the fire front at the surface quickly dries the biomass after it is laid down on the blaze. The massive heat released during pyrolysis is thus used as drying energy, and wet biomass with a water content of up to 50 percent can be carbonized. Once a high-energy fuel bed forms at the bottom of the Kon-Tiki, you can even pyrolyze freshly cut wood, leaves, or cattle dung. The Kon-Tiki thus works both as a dryer and a pyrolyzer.

Start by building an open stacked square chimney of dry wood in the middle of the kiln and about three-quarters of the kiln height. Ignite this airy wood chimney at the top with some tinder. Once the top two rows of the fire are burning well, it creates a train that pulls air down the walls of the kiln and back up through the middle of the wooden chimney. After about ten minutes, burning wood from the top of the chimney falls down the chimney and ignites the base. After another five minutes, the entire burning "chimney" can be collapsed and spread evenly on the bottom of the kiln.

Another 5 to 10 minutes later, a sufficiently hot bed of embers has been formed and the surface layer begins to be covered with white ash. This is the moment to add the first regular layer of biomass. Cover the glowing coals evenly but not too thickly. Once this new biomass layer also becomes coated with white ash, this is the sign that the feedstock has solidly reached pyrolysis temperature and exothermic pyrolysis will continue even in the absence of flaming combustion. It is time now to add the next layer of biomass. This will maintain a powerful flame front above the pyrolyzing material to consume down-convecting oxygen while combusting the smoke, thus protecting the char. This process is repeated for all the subsequent layers every 5 to 10 minutes until quenching. Consequently, working with the Kon-Tiki requires the constant presence of a person to add fresh biomass. If you wait too long, the char starts to oxidize, which reduces yield and increases the ash content of the biochar. Take care not to lay on too much, too fast, as this will weaken the flame, reducing its ability to capture the fumes and allowing smoke to escape.

Compared to an automated installation, the disadvantage of the Kon-Tiki kiln is that it must be hand-fed during the entire period of operation. Depending on the type, lumpiness, and water content of the feedstock, it takes two to eight hours to produce roughly 1 cubic meter of biochar in

the latest version of the Kon-Tiki kiln with side angles of
70 degrees. If you use dry woodchips, it only takes about
two hours; undried prunings take four to five hours; green
wood with logs, branches, and leaves takes up to eight hours.
Again, depending on the biomass, one person can operate
two to four kilns at the same time. On a working day, a person can thus produce with two to four kilns between 1 and
1.5 tons of biochar, which corresponds approximately to the
daily capacity (in 24-hour continuous operation) of a medium-sized industrial pyrolysis plant.

Another significant advantage of the Kon-Tiki is that the
biomass does not need to be homogenized, chopped, or even
pelletized but may simply be layered as coarse pieces up to
120 centimeters (47 inches) long. However, the charring time
is considerably longer than with dry, small biomasses. When
using fresh twigs and branches, the capacity of the Kon-Tiki
corresponds approximately to the amount of biomass that
accumulates in six hours of landscape maintenance or while
cutting firewood. Instead of tossing the branches and brush
unsuitable for firewood on a big pile that very slowly rots, or
is burned to mostly ash in a smoky fire, they can be charred
in the Kon-Tiki.

As the Kon-Tiki becomes full, make sure the last two to
three layers consist of only easily charred material, such as
thin branches or prunings, since larger pieces added in the
final stages will either remain incompletely charred or will
require too much time to burn, resulting in excessive ash
production.

Quenching can take place either from the top or the bottom. Quenching from the bottom works like this: About 20

minutes before the last layer is pyrolyzed, the water tap at the bottom of the Kon-Tiki is opened. Water flows slowly in from the bottom of the kiln. When the water meets the hot coals, it evaporates. The heated 600-to-700-degree-Celsius (1,112-to-1,292-degree Fahrenheit) water vapor rises through the char bed, not only making for a slow quench but partially activating the biochar at the same time. The hot steam serves to expel and react with condensates from the pores of the biochar. The biochar is thus cleaned, increasing the pore volume and the inner surfaces of the biochar. In this way, partially activated biochar is produced. Specific surface area measurement of various biochar produced with these methods ranges from 250 to 490 square meters per gram.

Alternatively, you can also completely douse the kiln from above, however, it would be to the detriment of the partial steam activation compared to the watering from below. The pore volume and the specific surface of the biochar would be smaller when doused from above. If you want to avoid wetting the char, so you can use it later, for example, as fuel charcoal, you can close the kiln, either with an airtight lid or simply with a thick layer of dirt to snuff it out and allow it to completely cool. The resulting dry quenched biochar is, however, much richer in condensates and also pollutants such as polycyclic aromatic hydrocarbons (PAHs). For fuel charcoal, this may be good, since the condensates and pollutants burn well, but for biochar used as animal feed, certainly not.

Biochar quenched with water generally fulfills all the requirements for the premium quality of the European Biochar Certificate (EBC). The open fire pyrolysis principle guarantees that the vast majority of the pyrolysis gas is expelled from the biochar and burned, not stuck on the bio-char surfaces and pores in the form of toxic condensates. The biochar is additionally cleaned and partially activated when slow quenched with water from the bottom. When quenched with water or liquid organic nutrients, the biochar becomes a highly efficient biochar fertilizer.

The pyrolysis temperature in the Kon-Tiki is 650 to 700 degrees Celsius (1,112 to 1,292 degrees Fahrenheit), with brief temperature peaks close to the flames going up to 750 to 800 degrees Celsius (1,382 to 1,472 degrees Fahrenheit). In this temperature range, the biomass, including its lignin, becomes completely charred. The result is a high-tempera-ture biochar of high quality, which is particularly suitable for animal feed, as a litter additive, for manure treatment, for composting, for drinking water filtration, for wastewater treatment, and generally to bind toxins and volatile nutri-ents. The Kon-Tiki biochar is less useful for direct application to soil, since it might adsorb labile soil nutrients and bind plant-signaling chemicals. Be sure to enhance biochar from the Kon-Tiki with nutrients before using it as a soil condi-tioner biochar.

Originally designed for agriculture in developing coun-tries, it is more and more apparent that the farmers of Europe, Australia, and America will also seize the chance to make their biochar themselves and use the Kon-Tiki to optimize their agricultural material cycles. The genius of the Kon-Tiki is in the elegance of the simple form and the avoidance of expensive moving parts and controls. Thus, the Kon-Tiki is robust and inexpensive. However, larger scale commer-cial and industrial biochar production require elaborate automation to reduce labor. This becomes far more of the cost than the basic reactor vessel and here the automated, continuously operated plants may remain unsurpassed. But for small and medium-sized farmers, landscapers, small winemakers (using their grape prunings), and gardeners who occasionally want to create their own high-quality biochar, there is no alternative that will be more efficient, less expensive, or as supremely beautiful.

By the end of 2015, farmers in more than 40 countries, from Australia, Indonesia, India, Nepal, Iran, Serbia, Sweden, Germany, England, Portugal, Ghana, Kenia, Namibia to the U.S., Mexico, Bolivia, Brazil, and Argentina, will have built Kon-Tiki kilns following the open source designs of the Ithaka Institute. In the second half of 2015, more biochar for agriculture was produced in Kon-Tiki kilns than in all industrial pyrolysis systems together. In Germany and Switzerland, plenty of small garden Kon-Tikis produce the biochar for climate gardeners.

INDUSTRIAL BIOCHAR PRODUCTION

NOWADAYS, BIOCHAR CAN be produced in modern highly controlled industrial pyrolysis units. One example, built by PYREG, is a continuous screw-fed pyrolysis machine that can annually transform 1,000 tons of biomass into 350 tons of biochar while generating 1,000 megawatt hours of heat. With this equipment, it is possible to permanently draw down roughly 1 ton of CO_2 from the atmosphere per 2 tons of green waste—this includes all the energy needed for transportation, shredding, and unit operation. The energy to heat the biomass to the necessary 400 to 750 degrees Celsius (752 to 1,382 degrees Fahrenheit) is generated by burning the accruing gases.

The PYREG company has, in the meantime, a good dozen units in Switzerland, Germany, and Austria, each with an annual capacity of 350 tons of biochar, which measured against the potential capacity for making terra preta is far too little. Another company that has developed industrial-scale biochar production technology is Pyrocal in Australia. Their "Big Char" continuous feed technology has been introduced in developing countries such as Vietnam. Many other companies are piloting different types of biochar production equipment at various scales, some of which can be used to produce electricity and heat.

CERTIFICATION PROTECTS AGAINST ABUSE

SOME ENVIRONMENTAL GROUPS fear unscrupulous profiteers might be tempted to clear wide swaths of forests to make biochar, bury it in the ground, and claim these measures would promote climatic protection. Against this backdrop, a group of international environmental organizations posted in 2009 a strongly worded warning against biochar. They claimed that biochar was "a new threat to people, land and ecosystems," and that it was a new form of "geo-engineering," a dangerous industrial-scale attempt to mitigate climate change that would require "many hundreds of millions of hectares of land for biomass production." They went on to explain that the aim of the "biochar lobbyists" in including biochar in the carbon trade could lead to genetically manipulated tree varieties being planted, felled, and carbonized.

HUMUS FORMATION IN THE ECO-REGION OF KAINDORF

THE AUSTRIAN AGRICULTURAL advisor Gerald Dunst was delighted: in November 2012, he was awarded the Austrian Climate Protection Award for his pyrolysis unit operating 24/7 and producing daily 1,000 kilograms (2,200 pounds) of biochar from cereal husks and paper mill sludge. His company, Sonnenerde, produces biochar and high-grade biochar-compost. The family business, founded in 1998 in Austria, has about a dozen employees and specializes in compost.

Sonnenerde is also a partner company to the eco-region of Kaindorf. Gerald Dunst has been advising more than 80 farmers on how to increase the humus content in their fields. Humus content has already increased to 6 percent in some cases since the introduction of compost, no-till farming, and crop rotation. Nowadays, those farmers no longer need to apply fertilizers or pesticides, as the soil's natural fertility inhibits pest infestation. Another benefit is that, unlike previously, heavy rain can be absorbed more efficiently and stored for future use.

The eco-region of Kaindorf, similarly distinguished by a number of environmental awards, was founded in 2007 under the chairmanship of Gerald's brother, Rainer. The six rural communities that make up the Kaindorf eco-region have set themselves the ambitious target of being CO_2 neutral by 2020 by using renewable energy, encouraging humus generation, and supporting terra preta field studies, among other activities. Since 2007, there have been seven symposia on these themes, and since 2013, the humus conference awarded a prize for the best humus farmers of Austria.

Additionally, the eco-region of Kaindorf launched a regional business in CO_2 certificates: carbon dioxide–emitting commercial enterprises make a contract with farmers who store carbon in the soil. An increasing number of companies in the region, including a supermarket chain and a brewing company, offset their CO_2 emissions by supporting regional humus enhancement. Using the proposed climate farming methods, a farmer in the Kaindorf region can bind the equivalent of 50 tons of CO_2 per hectare per year through humus regeneration. Through selling carbon credits to local companies at 30 euros per ton of CO_2, the farmer can earn up to 1,500 euros per hectare by sequestering carbon in his soil. In return, the farmers are committed to keeping the humus content in the soil stable for a minimum of 5 years.

Manfred Hohensinner, a former farmer and now the managing director of a local company that buys those carbon credits, enthusiastically endorses this new business model, which he describes as "a huge opportunity for the farming business, farmers as eco-warriors!" Another benefit of this carbon trading system is that the money for the offsets remains in the region. And the fertility of the soils is enhanced for generations.

Although the potential for abuse is always a possibility, there is, as we have previously mentioned, an enormous amount of underused organic waste freely available for carbonization, making the felling of forests for biochar production highly unlikely and economically absurd. Indeed, closed-loop biochar production, especially small-scale and on-farm systems, has often proven to be the most economically viable modes to date. In fact, augmenting humus using biochar supports the very causes many environmental groups so passionately care about: small-scale organic farming, renewable energy, and climate change mitigation.

It should be noted that directly sowing unamended biochar into the soil does not boost yields, at least in the short to mid term. Although adding biochar is likely to improve the water-holding capacity of a soil, without the enhancement of biochar with organic nutrients, the biochar has no major effect on plant growth. It is critical to first activate or "charge" the biochar before it is added to the soil. This can be done by mixing biochar with compost, or animal or human urine, or adding other microorganisms, minerals, and nutrients. The carbon sequestration potential of biochar is a welcome side effect, but it cannot pay the cost of biochar production, neither from free feedstocks, such as biowastes, nor from expensive forest wood. Biochar is economically viable as a soil amendment only when used as a carrier for nutrients, when it thus becomes a highly efficient multipurpose advanced fertilizer.

The danger of abuse can easily be countered by the certification of biochar regarding the quality of the product and its environmental compatibility. To this end, the European

Biochar Foundation has developed the European Biochar Certificate (EBC). Since 2012, European biochar producers can have their products and the sustainability of its production certified by an independent, government-accredited agency. Farmers or gardeners who buy such products should watch out for the appropriate seal. In North America, the International Biochar Initiative (IBI) has developed a similar certification program.

DO-IT-YOURSELF PYROLYSIS

ANYONE CAN MAKE their own biochar—from foliage, twigs, grass cuttings, coconut shells, rice husks, corncobs, reeds or canes, woodchips, and all sorts of other organic materials. The Kon-Tiki kilns are the easiest way to do it in your garden or on your farm. There are various other types of small and even mini pyrolysis units that can be used inside the house for heating or cooking. One type of unit is top-lit updraft gasifiers (TLUD), which are available in a variety of price brackets, or with a bit of skill, you can even build one yourself!

The basic design of TLUD gasifiers is as follows: the oven consists of two cylindrical containers with different diameters, one fitting inside the other. The inner cylinder is the pyrolysis chamber. The outer cylinder encloses the smaller inner cylinder, the top end being airtight. At the lower end of the outer cylinder, there are openings through which the outside air is drawn in for combustion. To make a simple version of this stove, you need nothing more than two metal cans, a drill, and a soldering iron.

To operate the stove, the inner cylinder is filled with relatively dry biomass (e.g., vine and tree prunings, agricultural residues, dry manure, and the like) and ignited. The use of tinder may help start the process. The pyrolysis gases in the inner chamber are drawn downwards by the airstream that flows upwards in the outer chamber. Gases flow through holes drilled into the bottom of the inner chamber and enter the outer chamber. In the outer chamber, the pyrolysis gases are mixed with air, rise above the combustibles, and again enter the inner chamber through top openings. At the top end of the inner chamber, the gases burn, after an initial smoky phase, with a clean flame and generally very little soot. As the biomass is burned cleanly, smoke inhalation, which in many developing counties is an all-too-frequent cause of death for women and children, isn't an issue.

There are numerous types of TLUDs commercially available worldwide, from camping stoves for environmentally aware trekkers (e.g., hobo stoves) to rice husk stoves in Asian countries. A method known as kuntan is used in Japan, whereby simple casks of rice husks are carbonized and then used to improve soil fertility. The Worldstove company deserves a special mention for providing TLUD gasifiers to the victims of the 2010 earthquake in Haiti. In Pune, India, the Appropriate Rural Technology Institute has developed a mobile TLUD gasifier suitable for preparing meals, which is fueled by pellets, woodchips, or small branches and costs around 24 euros. These stoves allow for up to one hour of cooking time per filling but can easily be replenished while in use. One kilogram (2.2 pounds) of dry biomass produces approximately 250 to 300 grams (0.5 pound) of biochar.

Daily use with a kilogram of dry biomass would produce in a year enough biochar to ensure long-term fertility for a 100-square-meter (1,000-square-foot) vegetable garden.

THE BENEFICIAL PROPERTIES OF BIOCHAR

BIOCHAR IS EXCEPTIONALLY porous and can have a surface area of more than 300 square meters (3,200 square feet) per gram, allowing it to absorb up to five times its own weight in water or other soluble nutrients. This latter property is known as adsorption capacity. Biochar is formed at temperatures between 400 and 750 degrees Celsius (752 to 1,382 degrees Fahrenheit) during pyrolysis, with higher temperatures typically generating higher surface area and thus improved adsorption capacity.

Cation-exchange capacity (CEC) is the measurement of a soil's ability to hold exchangeable ions and is used as a measurement of soil fertility. Typically, it is challenging to change a soil's CEC; however, biochar's electrostatic surface charges attract and hold positive ions (cations), making them available for plants and microorganisms. The capacity increases on contact with oxygen and soil particles and eventually reaches a peak. A high exchange capacity prevents the leaching of minerals and organic nutrients, ensures nutrient availability, and binds toxins harmful to soil organisms, rendering them unavailable to plants.

The sum of all these positive properties makes biochar an excellent carrier for nutrients. Microorganisms find an ideal habitat within the pores and are able to stimulate the soil, allowing a symbiosis between microorganisms and plant roots. Biochar acts like a sponge, sucking in water and nutrients, and then passing them on to plants when required. In addition to having a positive effect on climate protection, scientists have demonstrated the following benefits:

- Nutrients are better managed, leading in many cases to an increase in plant growth and improved yields using less fertilizer.
- Soil water-holding capacity is improved. Longer dry periods, a more common occurrence globally because of climate change, can be tolerated without affecting yields.
- Soil life is activated. Soil bacteria find an ideal habitat because of the high porosity surface area and release nutrients for plants. Mycorrhiza activities, a symbiotic association between fungi and the plant roots, increase, which means plants can better absorb water and minerals and can also offer improved defense against pests.
- Many different types of toxins or pollutants in the soil can be neutralized by the addition of biochar, including lead, cadmium, copper, and polycyclic aromatic hydrocarbons, pesticides.
- Soils are better aerated, allowing for improved penetration of air, water, and nutrients and plant roots to grow deeper and stronger.
- Soil-related methane and nitrous oxide emissions are reduced. This is especially interesting in terms of biochar's ability to reduce such emissions from rice paddies, which account for 12 percent of total global methane emissions.

However, we would be remiss if we did not strongly caution readers that not all biochar is the same and of equal high quality. If it is made from inferior or contaminated biomass, or under poorly controlled production conditions, the resulting biochar can contain toxins such as heavy metals, polycyclic aromatic hydrocarbons, and dioxins. Therefore, all biochar should be controlled and certified.

MICROORGANISMS: NATURE'S LITTLE HELPERS

IT IS IMPORTANT to understand that biochar is not a fertilizer but rather an excellent medium for storing nutrients, water, and microorganisms. As we have stated previously, before it can achieve its beneficial effects on the soil, it should be "charged" with nutrients and populated by microorganisms. This can be accomplished by using a variety of procedures, such as bokashi fermentation or composting.

Microorganisms are ubiquitous in nature and perform a myriad of beneficial services. They have existed on the planet for 3.5 billion years—bacteria and fungi were for a long time the first and only occupants of our planet, and they are involved in practically all natural processes. The enormous diversity of microorganisms drives the cycling of all organic matter. They can be found in all habitats and settle not only on the surface of living creatures but also deep inside. Without them and their metabolic products, life for plants, animals, and humans would be inconceivable.

Since industrialization, people's lifestyles have altered radically. The delicate balance between biotic communities in nature has been disrupted by, among other things, a false understanding of hygiene. Microorganisms are now often seen as a threat to be brought under control by the use of cleaning agents and disinfectants that increasingly find their way into the food supply. Unfortunately, but not surprisingly, various microbial species and functions are destroyed in the quest for this so-called hygiene.

In the early 1980s, Japanese agronomist and horticultural professor Teruo Higa developed a mixture of anaerobic (not dependent on oxygen) and aerobic (oxygen dependent) microorganisms in the course of his experiments on soil improvement. Spraying his mixture on soil and on plants, he improved the health of various fruits, vegetables, and rice. He termed his product "effective microorganisms" (EMS). His secret blend of some 80 species of microorganisms contains mostly lactic acid bacteria and yeasts. When applied under the absence of air, it introduces the lactic fermentation of organic matter, thus avoiding putrescence and uncontrolled decay. The function is based on the same widespread principle of nature that is also found in the guts of earthworms, the digestive tracts of humans and livestock, and in good humus soils.

In the mid-1990s, the use of effective microorganisms spread to North America and Europe. Since then, a network of producers, distributors, and consultancy agencies for EM products has been developed.

The probiotic microorganisms create via fermentation a milieu with strong regenerative, antioxidant, and energizing powers. Similar fermentation processes have been used for thousands of years for producing sauerkraut, sourdough, yoghurt, and the silage of fodder. In the process, microorganisms degrade complex organic matter into smaller molecules that are easy to assimilate by soil microbes and plants. In the conversion processes, there are no unwanted byproducts, such as ammonia, methane, or hydrogen sulfide, which make uncontrolled decay so bad smelling and negative for the climate. Using EM sprays in barns to treat the bedding and manure avoids the usual stench, improves animal health, increases the fertilizer value of the manure and augments its humus increasing capacities when applied to soil.

NATURAL FARMING—FARMING WITH NATIVE MICROORGANISMS

ON TOP OF the use of effective microorganisms, there are other ways of microbially influencing soil quality, such as the use of matured compost, plant extracts, compost teas, or sauerkraut juices. Many processes are employed, particularly in Asia, that are specifically aimed at promoting naturally available soil organisms. One example of this is the Korean natural farming system, which is famous in Asia but not yet well known elsewhere.

More than 50 years ago, South Korean farmer Han Kyu Cho began developing a form of sustainable farming using natural resources from the region. Cho achieved higher yields than conventional farming methods, without relying on synthetic fertilizers or pesticides, by actively promoting the diversity and interactions of soil organisms. It is a system without waste, without residual water, and without damaging emissions.

Cho's teachers were the Japanese farmers who had been intensively studying fermentation and needs-based nutrition using cultivated plants. Natural farming is based on precise observation of nature and is perfectly suited to small

EFFECTIVE MICROORGANISMS FROM CHIEMGAU

THE AGRICULTURAL CONSULTANT Christoph Fischer from the Upper Bavarian region of Chiemgau has been preoccupied with forms of sustainable farming since 1994. Concerned that agrocompanies were making traditionally independent farmers dependent on their patented seeds, he founded in 2006 the movement Aktion Zivilcourage (Action Civil Courage), one of the most successful German movements against genetic technology.

In the 1990s, Fischer started to experiment with Teruo Higa's effective microorganisms, whose range of effects became more and more impressive. He developed his own blending of microorganisms and created plenty of new products. His EMS not only increase soil fertility but also immunize animals and plants against some common diseases; neutralize the stench of manure and feces; can stabilize ponds; improve water quality; reduce liquid manure and sewage sludge; clean teeth; heal wounds; protect skin and hair; are good cleaning agents for the kitchen, tiles, and floors; and finally, diluted in water, can even make indoor plants happy.

The production site for the EM solutions is a huge stable fitted with solar panels. Inside, there are massive metal tanks, each containing a couple thousand liters of fermenting solutions. To produce the main blend, "EM-active," bacteria are fermented in airtight conditions for about ten days at a constant temperature of 40 degrees Celsius (104 degrees Fahrenheit) with a nutrient solution made from molasses. "At the beginning of the process the number of microorganisms doubles every twenty minutes," enthuses Fischer about his busy bacteria.

Microorganisms need nutrition, such as molasses, which is usually imported from countries like Paraguay or Indonesia. But Fischer wanted to be consistent with his philosophy of regional production methods, which is why he developed another nutrient solution, named "EM blond," in which the sugar sustenance for the microorganisms comes from local wheat and barley. Sugar beet molasses or honey can also be used for propagating microorganisms, though the composition, and thus the effects of the microorganisms alter with the various nutritional substances. Whether this change in quality is positive or negative depends on the intended purpose. Just like baking and cooking, there is plenty of room for experimentation—the only way to find out is to try it.

Fischer, however, is not content with just trading in EM products; he wants to educate the regional farmers and gardeners. He organizes seminars and advanced training courses. Every four to six weeks, he holds regional roundtable discussions, with 120 to 150 farmers in attendance, on biochar, lactic fermentation, manure treatment, and nutrient recycling. "We aren't the crowning glory of nature, we're just part of it. We could develop systems that conform to nature in order to create a paradise." Fischer is convinced, and he is finding increasing numbers of supporters; his Rosenheim project—developing sustainable farming—has attracted thousands of farmers and gardeners. By no means are they all organic farmers; in fact, the majority manage their businesses conventionally, or rather unconventionally conventionally—they work with mulch, mixed cultivation, composting, EMS, and Chiemgau black earth (see page 112).

Fischer addresses the whole spectrum of the local society. He lectures for Bavarian Trachtenverbände (clubs dressing in traditional regional costumes), local shooting clubs, and Catholic housewives, campaigning for his convictions and warning against genetic technology. Aktion Zivilcourage publicly states that members boycott genetically manipulated fodder and seeds and discontinue any form of cooperation with companies associated with or promoting genetic technology. The dairy cooperative Berchtesgadener Land only accepts milk from farmers who don't use genetically modified soy as feed. For dairy farmers, this involves additional costs of up to 10,000 euros per year, but they are convinced that it is the right thing to do. In the meantime, more than 150 rural districts and local authorities participate in the movement, with 4,500 signs at the entrance to villages announcing "Non-GM Farming." Groups with similar aims have sprung up in Hesse, Rhineland-Palatinate, Austria, and even in Egypt.

The secret of the movement's success is that it is as locally rooted as it is at home all over the world, just as down-to-earth as rebellious, just as close to the common people as it is visionary. Fischer looks out across the waters of the lake and is happy about the diversity that has emerged in the region. To his knowledge, nowhere else in the world is there such a density of committed farmers. Just as the microorganisms cooperate with each other, so do his Chiemgau farmers.

family businesses and urban garden plots. It can also help in the struggle against poverty and famine. Nowadays, the method is practiced in some 37 countries, most of them in Asia. Cho founded the Janong Natural Farming Institute near Seoul, which has trained around 18,000 people in this forward-looking method.

Natural farming cultivates indigenous microorganisms from healthy soils, onsite, propagating them selectively by fermentation. For users, this is considerably less expensive than commercial EM concentrates in which the microorganisms don't originate from the region. Those who choose natural farming, however, have to observe and take into account many processes and need a fair amount of experience and instinct to attain good yields. Asiatic gardeners and farmers have an advantage, because they can fall back on much more traditional experiences. Many methods of the fermentation of rice, soy, vegetables, or fish are in use there daily.

The first step in cultivating indigenous microorganisms (IMOS) is to bury a wooden case filled with boiled brown rice in healthy soil. The microorganisms living there then colonize the rice. Afterwards, the rice is mixed with brown sugar and water and fermented in an airtight clay vessel, allowing the microorganisms to multiply. Fermentation takes two to three weeks, depending on temperature. The IMOS are then available for a large number of uses. They can be further enriched with rice bran for the production of bokashi and compost. Kitchen wastes are added to the mixture and left to ferment in earth mounds with charcoal powder and soil. Finally, rice straw is placed on top as a covering to encourage incubation of fungi and bacteria that will stimulate fertility.

With this method, gardeners can build up a healthy humus-rich soil that doesn't overfertilize the plants but receives the required nutrients according to plant's needs, age, and maturity. To maintain the plant's health, you can spray them with a fermented mix of herbs, garlic, chili, and ginger, diluted with water. Eggshells doused in vinegar and bones provide calcium and phosphorus. Fermented fish wastes can also be used as valuable fertilizer. Human and animal feces, too, can and should be fermented with IMOS and returned to the natural cycle. In the bedding of chickens, pigs, and cattle the microbial solution is used to transform the animal's excrement into valuable manure, without stench.

The frenetic building activities before the 2008 Summer Olympic Games brought natural farming to Beijing. The overpowering smell of the large number of pigs brought in to provide meat for the countless workers led to vehement protests from the local population. The stench disappeared after introducing the methods of natural pig farming from South Korea. Since then, scientists at Renmin University in Beijing founded the Little Donkey Farm, which follows natural farming principles (see page 139).

STOCKHOLM BIOCHAR PROJECT

THE CITY OF Stockholm is in the process of investing in pyrolysis plants to produce both biochar and renewable energy from park and garden waste. The renewable energy will be turned into heat that can be added to a local or a district heating network. The biochar will be used within Stockholm in the public gardens. Some will also be given back to the citizens to be used in their private gardens and allotments.

In Stockholm, trees are planted in structured soils. Large rocks are compressed to create a stable structure, allowing the plant roots to spread even in environments with heavy traffic. A well allows storm water to enter the soil and gases to exit. The structured soils are made from local leftover construction materials, but the soil that is added is made from nonrenewable materials such as peat, sand, and clay. Stockholm started searching for a material that they could produce from local resources to substitute the finite materials. They stumbled upon biochar, and ever since 2009, biochar has been used in the city's plant beds. The improvement to the overall health of the trees has been remarkable.

Park and garden waste is collected from both the city and the residents in Stockholm, but it is difficult to dispose of. City managers and the energy company Fortum are focused on increasing renewable energy production. The combination of an interest in biochar, the difficulty in disposing of park and garden waste, and the goal of producing more renewable energy launched the Stockholm Biochar Project.

The vision of the Stockholm Biochar Project is to enable citizens and the city to tackle climate change while improving the local environment by producing biochar and renewable energy from park and garden waste. Using the biochar as a soil conditioner in public and private plant beds, the project will create a vast carbon sink that will help reduce overall greenhouse gas levels in the atmosphere.

CHAPTER 4

Ways of Producing
Terra Preta

Whether in compost heaps, kitchen bokashis, or slatted wooden crates—there are many ways of producing black earth. This chapter addresses many frequently asked questions and offers practical tips about the production of terra preta.

DEPENDING ON WHETHER you want to liven up a balcony, a small garden, or a large community garden—there are many paths that lead to terra preta production. Each hobby gardener, each urban gardener must choose the method that best suits their circumstances. The most important ingredients of black earth production are always the same—biochar, organic wastes, microorganisms, soil organisms, and a deep respect for nature.

The location and size of the garden, the amounts of waste matter and associated costs are all factors that will guide the choice of production. Do you want to order ready-made biochar on the Internet or make it yourself? Do you want to cultivate effective microorganisms or leave it to the professionals? Do you need to produce a lot or a little, slowly or quickly?

The question of whether you want to make use of your own excrement can make some people squeamish (hygiene is covered in chapter 6). Just to repeat quite clearly—this is an optional ingredient of terra preta. You can produce high-grade black earth without this component, simply with the addition of biochar and the fermentation of organic wastes. Those new to biochar production and use are advised to gain experience before incorporating urine or feces into their terra preta recipe. It is better to take things slowly than to reject the project because something has gone wrong. As with good wine, your own insights and experiences in making terra preta generally have to mature before you can reap the rewards that come with it.

People who have a garden right in front of their door can set up a relatively low-maintenance compost heap, fermenting fresh kitchen wastes there directly. A variation would be to produce kitchen bokashi, which can later be used on the compost. People with small or communal gardens farther away from their homes with some distance to their compost should use the bokashi method, adding it to the compost later. Those with balconies or patios should make use of wooden crates, as these take up less space. City dwellers gardening on soils that could be contaminated should also use crates set up as raised beds.

Large-scale production is feasible and necessary, especially for larger farms; you just need to scale up the mentioned variations with the help of sophisticated machinery.

More recently, farmers are enlisting the help of their livestock to add stable carbon to their soils. Dairy farmers in

Australia, for instance, have been adding between 150 and 300 grams of biochar per cow to the daily feed rations of their dairy herd since 2011. Not only has this proven to improve herd health and reduce manure odor, but the biochar can be found in the manure and is subsequently trampled into the pasture by the cows. Dung beetles and other soil fauna help to bring the biochar deeper into the soil profile, where it has been shown to substantially improve soil fertility and water management.

Nearly all organic waste can be used to produce terra preta—raw vegetables and seasoned and cooked leftovers, treated and untreated citrus fruits and banana peels; lawn cuttings, pine needles, and foliage; small twigs, wood shavings, sawdust, or straw; excrement from chickens, horses, pigs, cattle, and humans; the contents of litter trays—if the litter consists of sawdust or mineral bentonite; and meat remnants and bones. Kitchen waste can be composted in closed containers to keep out smaller animals. When using meat and other cooked food containing protein, care should be taken to blend it with raw vegetable wastes and cover the layers with biochar powder to reduce odors.

If your city has a food scrap collection program, cancel your organic waste bin and try to convince your friends and neighbors to do the same—organic waste is far too valuable as mulch or a soil amendment not to keep for yourself. While you're at it, you should also consider keeping yard waste instead. If you have a lot of trees on your land, then gather grass cuttings, weeds, garden waste, leaves, and twigs. Some of the fall foliage you can throw directly on the compost pile, and the twigs you can use to make into biochar.

THE HENNES' HENS, OR HOW TO
MAKE GOLD FROM CHICKEN MANURE

ON A FARM in Chiemgau, Bavaria, live the Hennes, their five children, and some 30,000 chickens laying nearly the same number of eggs every day. Conventional large-scale chicken operations traditionally come with foul-smelling coops housing stressed-out and nervous creatures pecking at each other. But here, everything is very different.

Each chicken coop houses 3,000 laying hens. They perch quietly on their roosts or cluck away in their screened nests, scratch around in the litter, go outside for a walk, and generally leave the impression of being satisfied. The chickens may be happier, as there is no stench of chicken droppings or the resulting byproduct, ammonia, which "makes the chickens nervous," according to Bernhard Hennes.

The Hennes farm is the first poultry business in Germany that in addition to producing eggs, turns chicken excrement into "black gold," homemade organic fertilizer made from fermented animal excrement and biochar called Chiemgau black earth—an excellent fertilizer with subtle smells of tobacco and vanilla. Marianne and Bernhard Hennes not only sell the black gold at their farm store alongside eggs, noodles, and honey, they also use it in their own flourishing vegetable garden.

Hennes was formerly a conventional chicken breeder until he took part, with 1,000 other farmers and horticulturists, in Christoph Fischer's Rosenheim project in which sustainable farming is practiced using mulch, crop rotation, composting and bokashi, biochar, EMS, and terra preta.

The Hennes' business, with some twelve employees packing the eggs and making noodles, isn't strictly an organic farm, but it is nearly chemical free. Five times a day, one-minute-long sprays obscure the coops with fine clouds of EMS. They stop the decaying processes and steer the biology in the henhouse towards fermentation. Fine-grained biochar is spread over the belt transporting the chicken excrement away, improving sanitation of the feces and neutralizing the stench. Hennes also uses biochar and EMS in the chicken feed and litter, and he is delighted that the veterinarian costs have decreased since he implemented this approach.

Other farmers process sewage sludge and grape pulp to make black earth substrates. Many let their cattle and pigs do the work in their stalls, where they mix the charcoal with fermented straw and their own excrement, simply by moving around.

Fischer claims that a mixture of 1 liter (0.25 gallon) of biochar and 1 liter of EM solution added per cubic meter of slurry can prevent significant ammonia emissions and almost immediately neutralize the stench. This factor appeals to many farmers, especially those who offer tours, farm stays, or other activities for visitors.

Huge amounts of chicken manure, straw, and biochar are dumped in a heap, blended, and then left to ferment under airtight plastic sheets. In a few weeks, the substrate is ready, "but it's still good after a year," explains Hennes. He spreads the black earth on his cornfields, which he shows off full of pride. Whereas his neighbor's plants are barely 30 centimeters (1 foot) in June, his own crops are chest high. "(Black) gold in the soil is better than gold in the bank," enthuses Hennes. "It's in a savings account where no one can find it."

1. COMPOST HEAPS

Ideally, a compost heap can be set up directly in the garden right outside your door—in an open wooden structure, or better yet a closed container, or even one that is open at ground level.

Shredded kitchen wastes and cooked leftovers that end up on the compost heap should be mixed with garden wastes and sprinkled with charcoal powder at a ratio of 10 to 1 (see page 142 for hints on making the powder). Finally, the residual organic wastes should be compressed for a few minutes, which can be done by using a ramming board or part of a wooden plank attached to a handle, or simply by stomping down on it with your feet. Lactic acid fermentation can only begin once most air pockets have been expelled.

Some small air pockets will inevitably remain, which will not really matter. In compost heaps, anaerobic processes (i.e., without oxygen), such as fermentation, and aerobic processes (i.e., with oxygen), such as humification by earthworms, happen concurrently. Nature knows no absolute states; even in forest soils there are localized areas of anaerobic and aerobic activities. Many microorganisms can alternate between the two milieus; in an aerobic environment they breathe, and in an anaerobic one they switch to another form of metabolism. There are some microorganisms that only exist by breathing and others that can only exist without oxygen, for instance, those producing lactic acid or alcohol, but when the organic material is tamped down, it is never so compact that all of the air is excluded. The soil organisms create ducts through feeding and burrowing, through which air can circulate, some of which gets inhaled by the oxygen-loving microorganisms.

After compression, lactic acid bacteria found in many vegetable wastes are responsible for kick-starting fermentation.

To accelerate the process, however, one can water the compost pile with an EM solution of diluted microorganisms. Half a cup of EM solution is sufficient for a 10-liter (2.5-gallon) watering can.

Anaerobic fermentation requires about one month to complete under ideal conditions; further aerobic humification takes between three and six months, depending on the time of year and temperature, so there is nothing left to do but wait.

Compost heaps should always be kept moist. During long continuous dry periods, they should be watered, just like plants in the garden. If they dry out, the microbiology comes to a standstill; if they are overwatered, there is a high risk of putrefaction and the leaching of nutrients. This is why you should always set up your compost pile in a sheltered area, for instance, in the shade of a tree, and cover it with breathable compost sheeting or boards as protection from rain. If rainwater cannot drain away, the pile can become soggy, which can lead to unwanted odors.

The best tip is simply to use your nose as a guide. Compost heaps should not stink; that is a sign of rotting and the release of methane and ammonia. If it does smell unpleasant, you can easily put a stop to it by neutralizing the odor with biochar. Further watering with diluted EM solutions may also be helpful.

2. KITCHEN BOKASHI WITH BIOCHAR

Bokashi is a Japanese term for fermented organic matter. Bokashi compost is made from compressed organic kitchen wastes that are placed in an airtight container to encourage lactic acid fermentation. The difference between traditional bokashi, which relies on effective microorganisms, and our process lies in coating the individual layers with moist biochar before compaction. Human feces, or humanure, can also be composted using the bokashi method (see chapter 6 for more information).

If you have purchased pre-mixed rock dust and ready-made terra preta, preparatory treatment of the biochar isn't necessary. If, however, you have made the biochar yourself, before use you should moisten it with 1 liter (0.25 gallon) of urine and, if available, 1 liter of EM solution per 20 liters (5 gallons), mixing in roughly 10 percent rock dust to the final volume. Then store the mixture in a closed bucket next to the bokashi container.

For the kitchen bokashi, or the humanure bokashi, you will need, in each case, two airtight containers, which, for ease of handling, shouldn't have a volume of more than 20 liters (5 gallons). While the first bucket is fermenting, the other is being filled. When you fill the buckets, regularly add some 10 to 20 percent biochar and spray it with EM solution. When the bucket is completely filled, compress the contents of the bucket as best you can and close it up airtight. The fermentation process in temperate climates requires this about once a month. Full buckets should be kept in a warm space at a temperature of 20 degrees Celsius (68 degrees Fahrenheit) or more. Longer storage times are not a problem.

Afterwards, the fully fermented material is worked into the compost in layers, where earthworms and other soil organisms mix it in and humify it within two to six months.

NOTES FROM A COMPOST DIARY

JUNE 2010: VERY curious about whether terra preta will change our 5-by-15-meter (50-by-160-foot) garden plot in Berlin. Our three mighty all-shading pines spread their roots so close to the surface that they immediately suck away everything that's nourishing. Even the flowers wither. Endowed with typical Berlin sandy soil, the garden has flatly refused to provide anything edible, apart from the annual bowl of sour red currants and five gooseberries. The carrots that I once planted were 1 to 2 centimeters (0.3 to 0.8 inch) long, and all other seedlings huffily died.

Now I want to turn our 60-by-60-centimeter (23-by-23-inch) compost heap, made by knocking together a couple of planks, into a black earth factory. During the week, I dump all kitchen and garden wastes onto it, and on the weekend, when I have more time, I add biochar at a ratio of about 1 to 10—sometimes a bit more. Then I trample it in. Every now and then, I water it with a watering can full of diluted EM (1 part EM solution to 100 parts water). Not strictly necessary, but better safe than sorry.

September 2010: We have constructed an oval raised bed some 120 by 60 by 50 centimeters (47 by 23 by 19 inches) from timber posts and lined it with a pond liner so that it won't rot over time. It is directly below the pines, where because of lack of sun, nothing has ever grown. First, we covered the bed with a layer of branches and twigs, then added a layer of our first finished terra preta soil from the compost.

Just below the topsoil of the compost, in the half-rotted section, there were an incredible number of red worm nests; deeper, there were fewer but many other creatures, such as springtails, wood lice, and the invisible hustlers and bustlers. The soil was black, warm, crumbly—a very different type of soil than the ones usually found here. All layers smelled mildly acidic and very good.

In the raised bed I have planted lovage, lettuce, arugula, rosemary, and oregano, and even a banana plant, which apparently can survive temperatures of minus 15 degrees Celsius (5 degrees Fahrenheit).

End of October 2010: Harvested lettuce and arugula—delicious! Without having sun or warmth.

November 2010: The compost heap is again full to the brim. Because I thought it was a waste to give all the good foliage from the surrounding trees to the Berlin waste disposal service, I dumped them on the compost pile—all in all, seven sackfuls— added biochar, then moistening the leaves with plenty of EM solution. My compost mentor claimed that by spring it will be at least half-rotted.

December 2010: Silent night, holy night, all is calm, all is bright round yon compost. Even when there's frost and snow, I tip my kitchen wastes onto the heap. The microorganisms overwinter in spore form but will reanimate as soon as it's warmer.

March 2011: The bananas have come to nothing. They froze— not really all that difficult to explain with temperatures reaching minus 20 degrees Celsius (minus 4 degrees Fahrenheit) for a number of nights.

May 2011: Great frustration with compost. I had imagined that in spring it would have turned to earth. I dug a large hole for a second bed. On skimming off the upper layer, I noticed there were far fewer worms than the first time and that in places it stank. The leaves did smell acidic, so they had properly fermented, but they were only half-rotted. My advisor thought I should just mix it with earth, put it in the newly prepared flowerbed, water it with EM, and in no time at all, it would fully rot. I did that and planted lettuce, peppers, and amaranth. On the first day, there were swarms of flies on the bed—a million flies can't be wrong.

End of June 2011: New attempt to clear the compost, this time successful. Wonderful soil with plenty of worms. Found two gigantic grubs. They weren't supersized tropical worms from the Amazon, as my son claimed, but merely compost-loving, old-wood-chomping larvae of the magnificent flower chafer.

August 2011: In the raised bed, the flowers are growing superbly. The lovage is so successful that we have to use the stuff in every soup and salad; in the meantime, we can't bear the sight of it. The cucumbers are colossal; the peppers, not. The sweet corn is growing well, as is the avocado I planted. The lettuces are fair to middling but still better than the ones in the terrace bed. Only the amaranth is proving stubborn, but then it is said they prefer poor soils.

May 2012: After a long winter, the lovage began to grow so rampantly that I've had to exile it to the terrace bed. Now it's sulking and visibly smaller, and instead, the borage that I planted at some stage has grown to a monstrous size. Terra preta is supposed to get better from year to year, and it seems as if my bed is trying to prove this. I now have a three-tiered agroforestry system—up above the pine tree, the larger plants in the raised bed, and below them the lettuces, which seem to prosper despite the lack of sun.

June 2012: The raised bed is running riot—like a jungle. Maybe because in May I gave it a kick start with some diluted urine (1 part urine to 10 parts water). Beans and peas with the tendrils climbing up the five strings that I have attached for them are growing at a rate of a few centimeters a day. The bed now looks like a sailboat stranded on the shores of the island of Utopia. Nearby, tomatoes, peppers, mulberries, the still-flourishing avocado plant that survived the winter indoors are growing, as well as lettuces and herbs. Edible chrysanthemums from last year have also reappeared.

Additionally, we've started another terra preta bed. It is 60 centimeters (23 inches) high and consists of three different elevations. We rubbed damp black earth into the wooden planks so that they don't look so new. Again, great joy on digging into the compost! Hordes of red worms wriggling around. I've stocked the two upper steps with branches and twigs so that the terra preta soil from the compost was just about enough. Then I sowed and planted and mulched with grass cuttings.

July 2012: Brrr! What a summer this has been. At first far too cold, then much too wet, every day thunderstorms. The freshly sown lettuces in the new bed were incensed and decided to remain underground, but the cucumbers and squash thrived. The lavish borage in the old raised bed, the leaves of which I used for salads blossomed more or less overnight and outed itself as a blanket flower (I was beginning to wonder why it didn't taste like borage). I vaguely remember last year scattering a few seeds on the bed without thinking. What an idiot! My only excuse is that the leaves are very similar, oval and velvety. Luckily, the family didn't suffer any poisoning!

August 2012: Picked beans, potatoes, tomatoes, peppers, lettuce and spinach. All in limited amounts, but still. The other beans that I planted in conventional beds are just beginning to blossom—so terra preta can accelerate harvest times by almost two months. Neighbors visiting my garden view it admiringly, perhaps even enviously.

3. STACKED SLATTED WOODEN CRATES

If you only have a small area available for gardening, stacked crates may be the best answer. Used wine crates or something similar measuring roughly 60 by 40 by 40 centimeters (23 by 15 by 15 inches), with slatted bases are a great option. At supermarkets, you can often get slightly damaged ones for free.

For each terra preta stack, you need two crates; in the upper crate, the plants grow in the matured biochar substrate, while in the lower one, new terra preta substrate is developing. First, in the lower crate put a layer of garden soil up to about the 10-centimeter (4-inch) level and on top of this a layer of biochar/bokashi, roughly the amount of a 20-liter (5-gallon) bokashi bucket. Care should be taken to leave about 5 centimeters (2 inches) on top that you can fill with regular soil, just in case there are residual odors from the bokashi. Afterwards, fill the stack to the brim with moist garden soil. Each crate should have about 50 percent bokashi and 50 percent garden soil. Watering with an EM solution at a ratio of 1 to 500 is optional.

Place the upper crate on the lower one and fill with a mixture of garden soil and mature compost, or if already available, mature biochar substrate (about half and half). Then you have a raised bed where you can grow your plants using high-density gardening techniques, such as square-foot gardening (see chapter 5, page 164). With this system, you can cultivate two plant types per crate. The crates should be watered regularly with EM solution, stinging nettle manure, or rainwater, as soil in crates dries out much quicker than in regular gardens.

Plant roots in the upper crate have plenty of space to develop and absorb nutrients as their roots descend into the lower crate, encouraging humification of the biochar/bokashi blend. A sign of quality is the presence of many soil organisms, particularly earthworms.

Once the upper crate has been harvested, the crates are exchanged—the former vegetable crate goes down below and the previous soil crate on top. Half the soil from the harvested crate is removed and replaced by a layer of biochar/bokashi and then filled to the brim with the soil you have just removed. The crate that was previously below, now full of wholesome humus, can be used for plants at the beginning of the gardening year. So the crates rotate every year, and the fertility of the soil in them, enriched by biochar, also improves year to year.

The terra preta stacks should be placed where plants grow well, where there is light and sun and where it is possible to practice "drop by" gardening. You can use your imagination when arranging them, make wonderful pyramids with them, or create a screen for privacy. Six to 10 stacks of crates are enough to recycle one person's kitchen wastes or, as the case may be, the products of your own metabolism.

4. FURTHER PROCESSING OF READY SUBSTRATES

The biochar substrates from the composts or bokashis are not yet terra preta as such; they are just the basis that stimulates soil life and initiates the optimum conditions for plant development. Proper terra preta can only form over many years, as the substrate continues to grow in symbiosis with the plant roots and soil organisms.

Mulching, rotating crops, and caring for the soil are not merely gardening chores but are an essential part of black earth production. You cannot keep the rich population of worms and microorganisms in your garden bed in the long term without providing continuous supplies of organic nutrients. Worms, especially, are big eaters; every night, they haul new leaves, stalks, or organic wastes below ground, and when this new pantry is empty, they move on to seek better gastronomic opportunities.

The various methods of mulching and crop rotation, briefly addressed here, are no different from those practiced in permaculture and many ancient cultures. Those wishing to reach a deeper understanding of such practices should obtain one of the many relevant books on this subject.

SOIL MOISTURE AND MULCHING

GENERALLY, TERRA PRETA soils don't need as frequent watering as normal soils, since biochar is capable of storing large amounts of water. However, climate gardeners must continually take care that the soil life, painstakingly developed, is not destroyed by wind and weather.

The ground should never be "naked"—in nature, one seldom sees naked earth, the exception being after recent natural disasters. The simplest way of safeguarding the soil with a protective covering is by mulching. Mulch keeps the soil moist and protects it from erosion from wind and weather. It also helps soften the effects of rain showers, as soils do not get as compacted if they are covered with mulch. Splashing is also reduced, so fruit bushes remain cleaner for picking.

In summer, mulch prevents the soil from drying out and from receiving too much solar radiation, at the same time protecting microorganisms from the damaging effects of uv radiation. In winter, the increased warmth under the mulch coating ensures prolonged life for soil organisms. Furthermore, mulch keeps earthworms and other soil creatures hearty and healthy, offering them a continuous supply of new nutrients.

Shredded organic wastes such as grass cuttings, straw, or leaves are well suited for mulching. You can leave unwanted weeds just where you plucked them, as, paradoxically, a covering of weeds protects against weeds. The mulch covering should always be just a few centimeters thick and loose enough to allow the rain to permeate; layers of dry material such as straw can be slightly thicker. Harvested beds should also be covered by mulch—they survive the winter better, and in springtime, new soil creatures coming to life have something to eat at the ready.

If a bed with mulch on it attracts too many snails or slugs, it is a sign the soil is putrid. Snails are biological indicators. Instead of combating them with chemicals or simply killing them, it might be better to ferment the mulch material, as snails definitely do not like this. Most of the time, they will withdraw in a huff.

REFRAIN FROM TURNING THE SOIL

EVEN IF GRANDMA did it and Grandpa hacked away at clods of earth with a heavy spade every fall after harvesting, digging up soil and plowing damages soil life. Today, many agricultural and scientific schools of thought—including conventional ones—agree on this.

A single thrust of a spade only 15 to 20 centimeters (6 to 8 inches) deep penetrates the humus layer. Soil organisms in deeper areas of low oxygen suddenly find themselves in the upper reaches, where exposed to air many die off quickly. Breathing soil microbes from the upper layers with access to oxygen end up in deeper layers, where they can suffocate. Plowing is often used to minimize compaction issues. However, soils enriched with terra preta are usually very crumbly, so they do not need loosening up. Not only does this mean less work for the farmer, but it is better for soil organisms and reduces soil-based greenhouse gas emissions.

CROP ROTATION AND MIXED CULTIVATION

EVEN TERRA PRETA can become impoverished if the same monocultures are cultivated on it over a long period. Soil exhaustion occurs not only because plants withdraw nutrients from the earth, but also because plant roots excrete various substances that accumulate with time, which can be detrimental at high levels. Both of these problems can be avoided by crop rotation and mixed cultivation.

Organic farmers are well versed in the benefits of perennial crop rotation. In three-year crop rotation, a plot is split into three parts. Plants requiring high intake of nutrients (e.g., potatoes, zucchinis, squash, celery, all cabbage species) are grown in one plot; plants with medium intakes (e.g., lettuce, spinach, kohlrabi, carrots, onions) in another; and plants with low intakes (e.g., beans, peas, herbs) in the third.

In the second year of cultivation, each of the plant groups is rotated to the neighboring plot so as not to exhaust the soil. The exceptions are: tomatoes, strawberries, and rhubarb, which should remain where they are.

In mixed cultivation, which is unavoidably practiced in square-foot gardening, you don't need to differentiate between high- and low-intake species. When using this method, you can plant cultures together that are compatible, whose root excretions are mutually beneficial, and whose scents and aromas ward off harmful insects. Carrots and onions, for instance, get along well, as do carrots and leeks, cucumbers and dill, beans and savory, tomatoes and parsley, celery and leeks.

Tomatoes and peas, potatoes and tomatoes, celery and lettuce, parsley and lettuce, cabbage and onions don't get along with each other. In garden beds, you can observe when growing together how they avoid each other.

Companion planting is based on a wealth of knowledge, experience, and personal observation, aspects that can only be touched upon here. Relevant books will help, and every gardener profits from insights gained from hands-on experimentation.

FAQS ABOUT TERRA PRETA PRODUCTION

How small should kitchen wastes be shredded?
Roughly the size of a thumbnail. The smaller the pieces, the quicker they can be digested. We also eat and digest in small quantities. The smaller the pieces, the better soil organisms can digest them.

Does this also apply to garden wastes? How small should they be?

Gardening wastes can be easily made smaller with a shovel. The material should be small enough to be easily compressed; otherwise, treat them in the same way as kitchen wastes.

Can I store garden wastes, such as leaves, in sacks and mix them gradually into the compost?

Yes, they ferment well in sealable sacks. The organic material doesn't go off and can be stored for long periods with minimal loss and used when needed.

How do you ferment foliage?

You can rake the leaves into a heap and shred them, if possible, by mowing them with a lawn mower. This stops the leaves from sticking together, and the damaged cells are easily colonized by microorganisms. Stuff the pretreated leaves into an airtight sack, and moisten them with an EM solution at a ratio of 1 to 500 for easily degradable leaves or 1 to 10 for oak and walnut leaves. The material should be damp but not saturated. Alternatively, you can add a beaker of live yoghurt culture diluted with water. Then press the air out of the sack, close the opening, and place the sack upside down in a sheltered location, preferably frost free. When conditions are frosty, the conversion processes cease, mostly reactivating once it gets warmer.

What about branches and wood? What size should they be?

Branches and wood should be finely chopped so that microorganisms have as large a surface area as possible to colonize.

Should you spray wood with urine when putting it on the compost heap?

During the transformation of wood, the activities of microorganisms are stimulated by additional nitrogen, so, yes, urine does help.

Is printer's ink or colored paper harmful?

Paper and cardboard should only be used if unprinted to prevent contaminants from accumulating in the soil.

Which buckets are suitable?

All non-corrosive, water-, and airtight buckets are fine, such as plastic buckets, diaper buckets, metal buckets with an enamel coating; otherwise, organic acids would corrode them. Ceramic, earthenware, or clay containers are all perfectly suitable. The containers shouldn't carry more than 20 liters (5 gallons) so that you can transport and empty them; any larger and they are difficult to manage.

Some (organic) mail-order firms sell bokashi buckets. Can you recommend them?

Yes. They are tougher than other buckets but also generally more expensive. They have a tap at the bottom to drain off liquids accumulated during fermentation. With biochar bokashi this isn't necessary. When using biochar, excess liquids don't lead to rotting processes.

How do you produce biochar?

There is an increasing number of methods, which are discussed in chapter 3.

Where can you buy biochar if you don't want to produce it yourself?

There is an increasing number of companies selling biochar. If possible, one should purchase biochar that has been certified either by the European Biochar Standard or the International Biochar Initiative standard to ensure that the biochar meets industry standards. Certified biochar can be ordered directly from the producers via the Internet. It is best to purchase biochar made as close to the purchaser as possible, as this will provide the best carbon-negative benefit. If there is no certified biochar available locally, it is important to at least understand what feedstock was used for the biochar production.

How do you make biochar fertilizer?

Mix biochar and urine at a volume ratio of 1 to 1. You may add some 5 percent of rock powder to further improve it.

What does the rock dust do?

Organic and mineral components are converted into tiny crumbs within the stomachs of earthworms. These minerals are introduced into the soil via rock dust, which is colonized by microorganisms unlocking the nutrients. Soils that have high lime or clay contents can do without rock dust additions. In this case, just apply urine-biochar close to the plant roots. If you are gardening on sandy soils, it is better to use rock dust.

Can you or should you use other ingredients?

If you have already gathered a fair amount of experience in fermentation processes, you can use other materials, such as coffee grounds, which are a good source of nitrogen, or saw-dust, which is a good source of carbon but has few nutrients.

Can you keep straw or should you always use fresh straw?
Straw can be stored for as long as you like. The already mixed stocks should last for one or two months. You will need it for layering the kitchen bokashi and, if you so choose, for feces.

Can barbecue charcoal be used, or does it contain too much toxic polycyclic aromatic hydrocarbons (PAHS)?
Proceed with caution: some cheap charcoal contains binders that may not be good for soil organisms. Also, production temperature plays an important role in producing soil amendment biochar. Generally, biochar is produced at higher temperatures than charcoal, allowing more volatiles to be burned off. When using barbecue charcoal, it is possible that certain volatiles may inhibit soil health.

What is the best size for the biochar pieces?
As the pieces have to pass through the stomachs of living creatures, particularly earthworms, they should be less than a centimeter (0.3 inch) thick. Biochar powder should contain a high proportion of powder, which, as a rule, is the result of abrasion during the production of biochar. A certain propor-tion of larger pieces, however, is not detrimental. The powder must always be moist; otherwise, it would be detrimental for your health and could be at risk of dust explosion.

What can you do with bones?
You can put small bones in the bokashi bucket. They also humify on the compost, though it takes longer. Larger bones

can be soaked in diluted vinegar for about six months so that they dissolve more rapidly and the phosphorus in them is released more easily. After preparing such a liquid, it can be diluted and used on the soil as fertilizer. You can also co-pyrolyze bones with other biomass; it will transform into nutrient-rich bone-char.

Can dogs smell bones in bokashis or on the compost and try to retrieve them?
Yes. If you want to avoid this, put the bones in a sealed compost container.

How long do effective microorganisms last? How do you keep them?
When kept in a cool place, EM solutions can survive for between one and six months. Once you have opened a bottle, you should ensure there is as little oxygen inside as possible. It is easier to squeeze out the air when using plastic bottles. The more contact that EMs have with oxygen, the quicker they change their composition. They consist of a mix that provides nutrients for each other, and over a longer period, only the lactic acid producers remain; the rest go into a passive state. As long as the smell doesn't change significantly, they can be used effectively.

Can high solar radiation or extreme cold kill microorganisms?
High UV radiation can harm or even kill microorganisms, which is why they shouldn't be exposed to sunlight for long periods. Bokashi buckets, however, are airtight and should be kept in warm places. At temperatures below 5 degrees Celsius (41 degrees Fahrenheit), microorganisms switch

to a dormant state. If the bucket freezes, it is not all that tragic; the biological processes slow down and will restart in warmer weather. The only danger here is that if the fermentation process hasn't already begun, it can lead to rotting upon thawing. If this happens, follow the guidelines below.

How warm should the fermenting location be?

The ideal temperature should be between 35 and 40 degrees Celsius (95 to 104 degrees Fahrenheit), but microbial activity also takes place at 20 degrees Celsius (68 degrees Fahrenheit), just slower. Boiler rooms are well suited. If a cellar or garage is the only option, providing temperatures of 10 to 15 degrees Celsius (50 to 59 degrees Fahrenheit), the process will simply take longer. Microorganisms need warmth to propagate.

How long does the fermentation process take?

With high temperatures, two weeks, but the rule of thumb in temperate climates is roughly four weeks. Fermentation provides the added benefit of conservation, since fermented organic material can be kept almost indefinitely.

How long do you need to ferment excrement?

Excrement should ferment for at least one month. The fermentation period is somewhat longer than for kitchen wastes, as it contains less sugar. Afterwards, fermented feces should humify together with kitchen and garden wastes on the compost for at least six months to a year. The proportion of feces should not exceed 30 percent. Once the substrate no longer smells bad, it can be used. So, once again, follow your nose!

Does the same apply to the contents of household pet litters?
Pet feces should not be worked directly into the soil. The hygienization and conversion to humus-like matter is the result of composting or humification. Once this process has been completed, it is safe to use as a soil amendment.

What kind of container should be used to store urine?
The container should be well sealable and non-corrosive. Plastic canisters or barrels with taps at the bottom are ideal. The best method, however, is to capture the urine directly in vats or buckets filled with biochar. Change the bucket before liquids appear on the surface.

How should a bokashi smell?
The proof of whether things are running smoothly and fermentation has taken place is in the smell. It should smell sweet-sour, like silage or sauerkraut. Unpleasant or even pungent smells are a sign that something has gone wrong and noxious fermentation has taken place. The same applies to foam or fermentation bubbles. Both are indicators of incomplete fermentation.

What do you do if a bokashi stinks?
Add biochar powder and, if you have them, effective microorganisms. Another option is to use yoghurt culture or hay infusions. Make sure to use yoghurt with live bacterial cultures. One container of yoghurt with live bacterial cultures is enough for about 10 liters (2.5 gallons). Compress and seal it again for another three weeks.

Does the same apply to the fermentation of human and animal feces?

Yes. A slight smell of excrement will remain, but it certainly shouldn't smell unpleasant. If it does, again, it is a sign that something has gone wrong. Additions of biochar and/or em will help.

What does a successful bokashi look like?

Another indication of quality is appearance. Fermented organic wastes don't change their original structure but should turn a yellowy brown. During the fermentation processes, a kind of white coating develops beneath the closed lid. It isn't mold but fungal hyphae, the threadlike cells of yeast that are active. This yeast is also the cause of the sweet-sour smell. If, however, there are patches of yellow or red, it is a sign that rotting is setting in, and then it will usually begin to smell bad. White meshwork with a large number of black spots is a sign of mold. If you see these signs, bury the material in the compost and don't use it directly for the soil.

When should kitchen bokashi be taken to the compost?

The contents of the bucket should only be put on the compost once the fermentation process is complete.

How often do you clean the bucket?

Preferably never. After emptying the bucket, you shouldn't rinse or wash it out; at most, dust it with biochar powder. The brim, however, should be wiped clean so that the lid can fit cleanly. The residues of the previous batch ensure the new batch is quickly colonized. This applies to both kitchen bokashis and humanure bokashis. The biochar and microorganisms take care of the cleaning duties. The upper parts of clay pots are particularly well suited for colonization by lactic acid bacteria. This saves work and is more hygienic than cleaning. Residues of cleaning fluids can contaminate and disrupt fermentation.

What should be done with kitchen wastes in winter?

In winter, you can continue producing bokashi in the kitchen or cellar. Lower temperatures, however, mean longer fermentation time. If fermentation grinds to a standstill because of frost, watch for signs of rot in spring.

What do you do with a compost heap in winter? Continue adding kitchen wastes and biochar?

Freezing conditions shouldn't have any effect at all. As soon as temperatures rise above 5 degrees Celsius (41 degrees Fahrenheit), the microorganisms reactivate and the fermentation process begins again. Alternatively, you can produce kitchen bokashi during this time.

After fermentation, do feces have to be mixed with bokashi and kept for a period of time, or can they go directly onto the compost?

The conversion to humus takes place together with kitchen bokashi on the compost. The excrement bucket provides certain raw materials, the kitchen bokashi the rest. Together, they end up on the compost heap or on the raised compost crate. On the compost heap, they have to be worked in layer by layer, roughly 50 percent bokashi, 50 percent garden soil or compost, and then trampled down. The worms then get to work and turn the material into humus.

Do you have to use more biochar and stamp down the compost once the bokashi and fermented excrement have been added?
Biochar only needs to be added at a volume ratio of 1 to 10 of the total volume. With feces and bokashi, more than 10 percent may be needed to contain odors. But, as usual, the material should be somewhat compressed. The aim is not hot composting, which converts organic material faster but also tends to suffer nutrient losses, but to get the worms working to create humus.

How moist should compost be?
The material shouldn't be too moist or too dry. This can be established with a simple hand test. Take a handful of mixed compost and squeeze it; no drops of fluid should appear and you should be able to form a ball that easily disintegrates when you push your thumb into it. If there aren't many worms in your pile or the humifying process takes too long, the reason is often a lack of moisture.

Is there an ideal blend of kitchen bokashi and garden waste?
Half and half is perfect, but usually you have more garden material than kitchen wastes or excrement. You should always take care that everything is well mixed. There really is no one best recipe; you merely have to adjust to the particular circumstances. If there is too much garden waste, you can store some in sacks for the time being or mulch it and use it as ground covering.

Can you leave ready, matured compost to its own devices?
Yes. Good compost is a living system. As long as it doesn't dry out, it can be left alone for a long time.

Are weed seeds killed during the fermentation process?
Fermentation severely inhibits the sprouting of seeds, but unlike hot composting, it does not kill them. So it is quite possible for unwelcome weeds to grow on terra preta beds.

Is there a minimum volume for a compost heap?
It should be at least 1 cubic meter (35 cubic feet), but ultimately, it will have to conform to individual circumstances and goals. There is a large difference depending on whether you are making terra preta for your window box, or you plan to supply an entire family with vegetables.

How do you use urine as fertilizer and how much do you need?
The rule of thumb is that one person's urine can fertilize and supply nutrients to an area of 300 square meters (3,200 square feet) per year. Urine contains nitrogen and significant amounts of potassium, as well as valuable micronutrients. The most efficient method is to soak the urine into biochar and to apply the urine-biochar into the root zone. If you wish to fertilize plants directly with urine, it must be diluted with water at a ratio of 1 to 10, as it would be toxic when used in a more concentrated form. Plants only require a lot of nitrogen in the growing phase, so after August (in the northern hemisphere) it isn't necessary, the exceptions being lettuce or vegetables that can be harvested into fall.

How do you know whether plants need nitrogen or whether they already have had enough?

If the leaves are light green, it is generally a sign of lack of nitrogen. If the plants grow too quickly, have a high leaf mass with few blossoms, or are not particularly steady, then the chances are that they have absorbed too much nitrogen. If this is the case, they may be more susceptible to infestation by harmful insects. In the blossoming phase, plants need more phosphorus to produce fruit. Phosphorus is present in compost and manure.

First and foremost, soil health should be encouraged instead of focusing more specifically on fertilizing specific plants. Soil organisms, which result from healthy soils, ensure plants receive the nutrients they need.

You too can join the cycle—healthy soils produce healthy plants and healthy nutrition for all people, who in return for their metabolic byproducts, provide healthy soils. Just as humans desire to enjoy good food and a variety of nourishments, so too do soil organisms and plants.

COMMUNITY-SUPPORTED AGRICULTURE: LITTLE DONKEY FARM, OUTSIDE BEIJING

THE 15-HECTARE (37-ACRE) Little Donkey Farm is situated just northwest of Beijing, in the foothills of the Fenghuangling mountains, and is the first community-supported agriculture (CSA) project in China. Eggplants, green beans, lettuce, napa cabbage, corn, herbs, and many more crops all grow on small parcels of land. Other vegetables are cultivated in traditional greenhouses heated by the sun. In the farm store, one can purchase all the healthy products grown there, and yes, of course, there is a little donkey.

Pigs, goats, and chickens roam around on a thick layer consisting of bokashi and straw, trampling and mixing in their own excrement. The urine is absorbed, and the microorganisms ferment the mixture below the surface. During the year, an excellent fertilizer is produced for the vegetable plots and greenhouses.

On Little Donkey Farm, the operators—Renmin University and the Haidian District council—have tried to find new ways of combining urban farming with the traditional farmer's knowledge of Asian small-scale agriculture. Directors, Shi Yan and Professor Wen Tiejun, from the Institute of Advanced Studies for Sustainability, teach skills and traditional farming techniques—permaculture systems and the natural farming methods of Han Kyu Cho, founder and director of the Janong Natural Farming Institute in South Korea (see page 98). China has become the largest consumer of fertilizers and pesticides in the world, but on this small farm, they are banned.

Community-supported agriculture is a growing and promising trend based on community, risk sharing, and trust. Consumers pay an annual fee for a share in the farm's harvest, and producers supply at regular intervals fresh, healthy produce that is harvested over the growing season. In 2011, 430 families paid an annual subscription to Little Donkey Farm and in return received a share of the harvest. A further 260 families leased small plots of land on the farm to grow their own vegetables. Gardeners and farmers on the staff organize the appointment of the plots. The farm has a permanent staff of 25, with 10 trainees at any given time.

The concept for the farm comes from, among other places, Earthrise Farm, a similar project in western Minnesota where Shi Yan worked for six months. Since the organic farmer Robyn Van En coined the term "community-supported agriculture" in 1986, more than 13,000 CSA networks have sprung up in North America. The advantages for both sides are that the farmers have at their disposal operating capital for organic production methods and are far less dependent on banks for financing. CSA customers know where the crops come from and in some cases can provide input as to which crops are cultivated.

Little Donkey Farm is more than just a CSA, however; it is also an education center and research station of Renmin University, which is used to evaluate sustainable agriculture methods and techniques. The project also works with many NGOs, including the international network PeaceWomen Across the Globe. This network of 1,000 PeaceWomen, which was nominated for the Nobel Peace Prize in 2005, works globally for peace, social justice, and lifestyles that conserve resources.

In November 2010, the Chinese coordinator of PeaceWomen invited Haiko Pieplow to an international conference in Beijing on sustainable farming. Together with Shi Yan, a terra preta workshop was held at Little Donkey Farm with more than 50 men and women participating from more than 10 Asian countries, including Han Kyu Cho. As the farm practices the principles of natural farming, the materials and microorganisms necessary for producing terra preta were already in place, though large-scale production of terra preta substrates had not yet managed to become established.

Shi Yan loves telling the story of the donkey, the symbol of Little Donkey Farm. A number of years ago there were heated discussions about what the future of urban, organic farming would look like. One side was saying donkeys had to be used to plow the field. The other side was saying we need tractors. Professor Wen had the final word: "Let's stop discussing and just try it out. I will buy a donkey, and my wife will name it!" His wife named the donkey Professor, because they couldn't find anyone who still knew how to work a donkey, so it was left to the donkey to teach the farmers how to work with it!

How to make
≡ TERRA PRETA ≡

Collecting bucket with lid
(10 L/2.6 gal.)

Charcoal bucket with lid
(2–5 L/0.5–1.3 gal.)

Charcoal powder
(1 L/4 c.)

Handful of microorganisms
(30 g/1 oz.)

2 plastic mesh boxes
h20 cm x w36 cm x d26 cm
(h8 in x w14 in x d10 in)

Wood ash
(optional)

Soil
(20 L/44 lbs.)

Compost
(10 L/22 lbs.)

Compost worms
(*Eisenia fetida*)

1 Make charcoal mixture. Put charcoal powder, microorganisms, and water (or urine) into the charcoal bucket. Mix them well and close the lid so that it will not dry out. Optionally, a little bit of wood ash could be added, too.

2 Put kitchen waste into the collecting bucket and cover it with charcoal mixture. You could add cooked material—also, citrus fruits. You could add meat, bone, or fish, but composting takes time and may attract vermin! Close the lid.

3 Each time you throw kitchen waste away, cover it with charcoal mixture. Make layers of kitchen waste and charcoal alternately. If you like, chicken manure or other animal manure could be added, too.

4 When the bucket becomes full, cover the surface with charcoal mixture and close the lid tightly. Wait four weeks. During this period, microorganisms start fermentation.

How to make
=== TERRA PRETA ===

5

After four weeks, put 5 L (11 lbs.) of soil in the bottom of the plastic box and add fermented kitchen waste in the middle of the box.

6

Cover the box fully with another 5 L (11 lbs.) of soil. Add compost worms. This box will be the lower box of the two-tier system. Fill the box fully so that there is no gap between the two boxes!

7

Install another box on top of the lower box and add normal soil and compost (ratio is 1:1). The bottom of the top box should be open. Then you can start planting!

8

Roots know where they should grow. When the terra preta compost in the lower box is ready, they start to use the nutrients from there, too.

9

The next year, you can swap the boxes around. In the lower box, you can prepare terra preta compost for next year: add fermented kitchen waste (repeat steps 1 to 6).

10

Optionally, if you have a garden, you can make a raised bed directly on the ground!

(The idea was suggested by Dr. Haiko Pieplow in Germany.)

Take care that there is no gap between the two boxes and start planting again!

CHAPTER 5

Biological and Horticultural Diversity

Whether in urban or rural settings, private or communal gardens, window boxes or square-foot gardens, homemade terra preta is teeming with life.

IN THE BIBLE, paradise is a luxurious garden—the Garden of Eden. However, the concept of finding happiness in a garden in which milk and honey and love flow, where flowers never stop blossoming and mouth-watering fruits are always available, is even older than the Bible and is shared by many religions and cultures. In Sumerian, this place is called *Guan Eden*, or the fringes of the celestial steppes. In Genesis, the Garden of Eden was irrigated by four rivers; later scholars and religious leaders identified them as the Tigris and Euphrates, Indus, Ganges, and Danube or Nile and, depending on who you believed, located the lost paradise in Mesopotamia, Azerbaijan, Bahrain, Ethiopia, or Jackson County, Indiana.

Jews, Christians, and Muslims were taught to believe that Adam and Eve were expelled from paradise for disobeying God.

But maybe, after tasting the forbidden fruit of the Tree of Knowledge, they had the task of introducing this experience to the rest of the world so that new paradises could be created. The Hebrew name for Adam is derived from *adamah*, meaning earth or ground. The biblical story of expulsion from the Garden of Eden can also be interpreted as a symbolic history of humankind—hunter-gatherers getting their sustenance from the wealth of nature become farmer-pastoralists earning their bread with the sweat of their brow. As Cain, the farmer, and Abel, the shepherd, demonstrated, things can sometimes go very wrong.

The primeval human yearning for nature's bounty has endured, as even today, many people seek to create a natural paradise on earth. One example of many is the oldest German cooperative garden, founded in 1893 in Oranienburg, Berlin, which is still in existence today. The 18 founders were vegetarians and, in the face of industrialization, propagated back-to-nature, holistic, and non-alienated lifestyles. This Garden of Eden had to be laboriously re-created from the barren and sandy soil of Brandenburg. The soil was reinvigorated by many hundreds of tons of horse droppings that the settlers gathered from the roads and worked into the soil.

GLOBAL BUTTERFLY MOVEMENTS

ARTIST KRISTINA BUCH created a totally different example of a Garden of Eden, one for butterflies. She planted roughly 3,000 plants, including 180 species on a 10-by-10-meter (32-by-32-foot) plot, which served as forage crops for 40 different species of butterflies. Countless butterflies—ancient symbols of charm and love, of freedom and transformation—hatched during the 100-day exhibition in her open-air wilderness demonstration.

Butterflies have an astonishing capability. At a particular stage of development, the caterpillars overeat—far more than their metabolism requires. During this process, they begin to die, but a select few cells are then activated; these are the imaginal cells. They cluster together, first in small groups, then in somewhat larger ones, passing information from one to the other by resonating. They are completely different cells from caterpillar cells, and the British morphologist Don Williamson believes all types of organisms that produce larvae originate from a totally different species that merged by chance at a more recent stage in evolution. Although the caterpillar attacks these alien cells, it cannot destroy them quickly enough; the dying caterpillar body provides sustenance to its newly emergent form. At some stage, the cluster of imaginal cells reaches a critical mass.

What then happens is most poetically described by Norie Huddle in her book *Butterfly*. "Then after a while, the entire long string of imaginal cells suddenly realizes all together that it is Something Different from the caterpillar. Something New! Something Wonderful! ... And in that realization is the shout of the birth of the butterfly! ... Each new butterfly cell can take on a different job ... There is something for everyone to do. And everyone is important. And each cell begins to do just that very thing it is most drawn to do. And every other cell encourages it to do just that. A great way to organize a butterfly!"

Wouldn't it be great to see a similar type of metamorphosis happen to change our current economic paradigm of "take-make-waste" into something more circular, more nourishing? Imagine that the caterpillar symbolizes the ravenous nature of old-school capitalism; the imaginal cells are the global union of those searching for an alternative, who want to practice urban gardening, prevent climate chaos, share instead of own, and cooperate instead of compete. The butterfly symbolizes the new economy of solidarity and freedom that satisfies human needs, particularly by means of local economies and flourishing gardens in urban, suburban, and rural environments.

FROM URBAN GARDENING TO TRANSITION TOWNS

EVEN IF YOU don't share this vision, it is difficult to deny that urban or suburban gardens are an important ingredient for self-sufficiency. New gardeners discover primordial pleasures, such as sowing, planting, nurturing, nursing, harvesting, and preserving. They learn new skills that connect with nature and how to taste what they have sown with their own hands. This brings a deep and abiding satisfaction that cannot be found when consuming industrially produced products.

To be sure, urban gardening is nothing new. In the slums and megacities of the world, small and tiny gardens are sometimes a question of survival. In Lubumbashi, for instance, the second-largest city in the war-torn and hunger-stricken Democratic Republic of the Congo, some 60,000 tons of vegetables are harvested annually in urban gardens. The floating gardens of Mexico City, mentioned on page 51, also provide sustenance for many urban dwellers. In Havana, Cuba, communal gardens supply more than 70 percent of the vegetables consumed in the city—perhaps an urban world record. After the collapse of the Soviet Union, which resulted in diminished supplies of oil and fertilizer, Cuban farmers were forced to go in a completely new direction and developed some of the first agricultural production not based on fossil fuel in the world. Out of necessity and conviction, these methods were organic, as they had no access to chemicals and fertilizers.

It is inspiring to see how much urban gardening has grown in recent decades. Guerrilla gardening originated in London and New York—this is when people or neighborhood associations take over derelict land sites and transform them into gardens.

In Totnes, England, environmental visionary Rob Hopkins founded the transition town movement, which has spread to the U.S. and beyond. Its objective was to prepare cities, towns, and boroughs for an oil-free era and make them resilient to the anticipated economic and ecological shocks that are expected to occur as a result of climate change. Residents work on communal plans for transitioning from fossil fuels to renewable energy sources, promote construction and the use of bike paths, set up bartering schemes and regional currencies, and plant community gardens, fruit and nut trees, or transform parking lots to vegetable plots.

Hopkins was inspired by permaculture, a concept developed in the mid-1970s by Australians Bill Mollison and David Holmgren for creating sustainable cycles in harmony with nature. Originally, permaculture focused only on agriculture but has since expanded to encompass town and country planning, decentralized energy supplies, and the establishment of social structures. For Hopkins, permaculture thinking was like wearing "green glasses": "Suddenly, you no longer see problems but just solutions." The concept of converting various waste streams into terra preta aligns well with the underlying philosophy of permaculture and transition towns.

Many people who practice urban gardening are at the same time explicitly political and explicitly unconventional thinkers. They link political statements against the agro-industry to private feel-good strategies. They no longer want to drop out of the system and look for a sanctuary in the country, while experimenting with ways to improve living in the city. They do their gardening individually or communally; cook organically, seasonally, and locally; are vegetarians or not; they reject, more or less strictly, the exploitation of animals, as well as farmers in developing countries.

They are mobile and extremely well networked and exchange information via the Internet or smartphones. Instances of young gardeners proudly posting their home-grown vegetables on social media before polishing them off are common. The young get on famously with older gardeners, who often help them with bringing up their children and who have experienced the joy of watching something grow, blossom, and prosper under their care and now want to try it out on plants.

PERMACULTURE AT THE FARM

BY ALBERT BATES, author of *The Biochar Solution: Carbon Farming and Climate*

FOR THE PAST 20 years or so, the land management decisions of the Farm (a 45-year-old intentional community in rural Tennessee, with a population of around 200) have been informed by permaculture. Many people serving on various village committees, as well as some in public office in the surrounding county, have permaculture design certificates awarded after attending courses at the Farm.

A few years ago, the Land Use Committee began getting interested in preparing for climate change in the coming century. The goal is to avoid the mistakes of the past, such as planting heirloom tree species from earlier centuries while climate isotherms are migrating poleward 70 miles per decade.

"Don't put in dams until you've keylined the place," Darren Doherty advised. The keyline method was developed by Australian stockman P.A. Yeomans in the 1950s. Yeomans's plow has wings on the sides of the shovel so that during cultivation, the soil is gently raised and loosened without turning a furrow. Rain and air enter and release the minerals that chelate and loosely attach themselves to clay particles and humic acid. Immediate results are dramatic. Accumulations of 400 to 600 tons of topsoil per acre each year are possible. Keylining can annually deepen topsoil 10 to 15 centimeters (4 to 6 inches) and darken it to a meter deep in less than a decade.

Each year since 2010, community members have been reclaiming more and more of the pastures and croplands of the

Farm, using a combination of the keyline method, remineralization, compost tea, and biochar. The biochar is prepared using an earthen pit and dried bamboo, quenched with urine, then pulverized in a leaf shredder and composted for several months before application. Teas are brewed in 1,000-gallon batches and applied as a slurry, along with biochar and minerals, as the keyline plow moves through the field applying roughly 30 gallons per acre. The teas contain kelp, humates, folic acid, fish oil emulsion, bat guano, feather mean, virgin forest soil, deep pasture topsoil, composted animal manure, composted kitchen scraps, composted poultry litter, worm castings and liquor, and finely ground biochar.

The kelp, fish oil, and most of the composts provide rich food for the microbes while they brew. The humates are million-year-old deposits with diverse paleobacteria. The bat guano is drawn from distant caves rich in trace minerals and packed with still more varieties of exotic bacteria. The two kinds of soil contain a complex of two discrete living microbiomes, one the fungally rich virgin forest and the other the bacterially dominated grasslands. The fine biochar particulates provide enough soil structure to retain water—about 10 times the volume of the biochar itself—and aerobic conditions, while providing a coral reef–like microbial habitat. The animal manures, worm castings, feather meal, and compostables all contribute to the biodiversity of available microfauna.

This process already proved its worth during a three-month summer drought. The inhabitants of the Farm are building tilth, recharging their aquifers and adding water-retaining capacity to their soils. As the Farm gradually becomes more tropical, this capacity will be critical for keeping the rain that falls in the wet times of year.

DIVERSITY WITH A CAPITAL *D*

SOCIAL AND (MICRO)BIOLOGICAL diversity are key concepts in gardening. Gardens thrive on diversity, on the abundance and exuberance of a wide variety of different living organisms. The beauty of terra preta is that it encourages all of these. Black earth increases the number and variety of microorganisms in the soil and plant species that grow on it. Indirectly, it benefits both animals and humans, who gain nutrients from it and give back to the soil when needed. Black earth promotes social diversity of the garden as well, because it is something that nearly everyone can make on their own. It can be created in villages and cities, in intercultural community and communal gardens, in cloisters and schools, greenhouses, botanical gardens, forest gardens, or on farms of many different sizes.

An increasing number of initiatives around the world are transforming the old adage "Another world is possible" to "Another world is plantable." They are sowing non-hybrid seeds so that gardeners can remain independent of hybrid or GM seeds promulgated by Big Ag.

Other groups practice community-supported agriculture (see chapter 4, page 139), or form urban initiatives to revive wastelands in shrinking cities, or map out wild fruits for public consumption, or grow food with refugees.

The dramatic decline in global biodiversity is quite possibly more threatening than climate change to the future of humanity. Creating spaces where biodiversity can thrive should, therefore, be a key goal in the broader context of creating terra preta soils.

If every person were to create at least 2 square meters (20 square feet) of garden, at home, school, or work—perhaps in the form of a window box or simply by using some free space—that would equate to naturalizing an area of almost a million square meters (10 million square feet). This could provide vast new areas of refuge for endangered beneficial fauna such as butterflies and bees, while at the same time providing new reservoirs for 20,000 tons of CO_2 equivalent. Everyone can do something to protect and promote biodiversity. Perhaps the most beautiful and simplest option is by way of building a garden in a window box.

WINDOW BOXES WITH TERRA PRETA

WINDOW BOXES CAN be placed on practically any windowsill; it doesn't matter whether the apartment is facing south and on the top floor or facing north and on the ground floor. All that is needed is five planks of wood cut to the desired size, which can be done at a DIY store if you don't have the necessary equipment at home. With wood screws and a cordless screwdriver, you can have it mounted on the sill in a mere 15 minutes. The window box should be at least 15 to 20 centimeters (6 to 8 inches) deep to allow space for the roots to grow. If the windowsill isn't wide enough, you could extend the box beyond the sill, as long as more than half of it remains on the sill and it is well anchored to both the window frame and the sill.

An average windowsill can host a 0.5-square-meter (5-square-foot) garden. Four window boxes filled with the right type of soils should be enough to provide lettuce and herbs for nine months and, every now and then, radishes, parsley, a few tomatoes, peppers, eggplants, flowers, and for the sheer joy of it, one meal of potatoes.

The object of window boxes isn't to provide a substantial portion of your daily diet but simply to enjoy watching your own food grow day by day—to see how nature finds space in the smallest of niches, to gain satisfaction from taking care of a tiny bit of nature, and to make way for a verdant wedge amid the uniformity and monotony of concrete. It is the joy of experimenting with rare species or the simple pleasure of looking out the window at something green. Bees drop in, butterflies land, and ladybugs delight in the few aphids they discover. Two square meters (20 square feet) offers the possibility of changing your perspective on everyday life. The comparatively modest size means that maintaining it isn't arduous. Even children seem to enjoy the simple pleasure of gardening and can independently care for their own window box while learning about how plants grow. After only one harvest season, they will no longer view produce as the odorless, sterile vegetables normally found on supermarket shelves.

The soil in your window box should always be covered with leaves or dried plant wastes, or even your own kitchen wastes, to prevent against drying out and to reintroduce nutrients into the soil. It is also a good idea to place compost worms directly in the soil of window boxes, where they convert organic wastes to humus and encourage biological activities.

As window boxes are in relatively sheltered positions, they profit from heated apartments during late fall and early spring, so in many climates, people can enjoy greenery throughout a large part of the year.

PRINZESSINNENGARTEN, BERLIN

Guerrilla Gardening with the Green Che

THE NOISE OF cars drifts over Prinzessinnenstrasse, from which the project gets its name, reminding us that Prinzessinnengarten lies in the middle of the built-up area of Berlin, Kreuzberg. But the derelict area around Moritzplatz, which for decades lay in the shadows of the Berlin Wall, has become a green paradise over the past several years. Project pioneers Robert Shaw (35) and Marco Clausen (39) cleared the rubble from the wasteland and, with friends and neighbors, set up a 6,000-square-meter (65,000-square-foot) communal garden that is visited by nearly 50,000 people a year.

Robert Shaw, beneath his beret, looks astonishingly similar to the young Che Guevara, especially when he laughs, with just his blue eyes dispelling the comparison. However, ideologically the two are far from similàr; the guerrilla gardener is decidedly easygoing. He cares not for overturning society but rather for overturning sacks, preferably those filled with produce. When harvesting potatoes, which grow in recycled rice or flour sacks, "we just need to turn the sacks upside down," he quips.

The city leases the land but reserves the right to sell it. Therefore, Shaw and Clausen developed the concept of a mobile organic garden so that, in the event of an eviction, the garden could be moved elsewhere easily. Nomadic Green, the name of their non-profit company, works with youth; organizes workshops, concerts, and events; and helps provide new opportunities and perspectives to welfare-to-work participants. The nomadic nature of the garden is so because the vegetable beds are made of recycled rice and flour sacks or wine crates and are fully portable. It's a smart concept that attracts journalists and visitors from around the world, ensuring various awards are heaped on the garden. Thirteen full-time jobs have been created; affiliated gardens have been established in 16 schools, universities, daycare centers, and backyards; and more than 100 volunteers put in around 30,000 hours of garden work per season.

The lease for Prinzessinnengarten, however, is due to run out anytime. Investors are standing by to build commercial properties on the ground, or at least on a portion of it. In the meantime, the garden can no longer be relocated; it is deeply rooted in the neighborhood and is based on "social connections within the community," counters Clausen. It is certainly very odd: reporters from Argentina to China to New Zealand write articles about the project, the city council of Seoul makes enquiries about how the system works, yet Berlin city administrators find it hard to promote the long-term existence of one of the most famous urban community gardens in the world.

Who knows, maybe the plants will soon start demonstrating; even they are more rebellious than elsewhere. Every fall, you see them going on the long march to their winter quarters—in wheelbarrows and in bicycle baskets transported by garden volunteers.

Come summertime, the garden is once again swarming with visitors and hobby gardeners. In a converted ship container, they order their Cuba libres, coffees, or fruit juices and eat pizzas topped with homegrown vegetables. No meat dishes are to be found on the menu. The "Green Che," Robert Shaw, lounges in a chair beneath a locust tree and explains how urban agriculture in Cuba inspired him. In 1999, he lived as a film student in Santa Clara, which was seized by the real Che during the Cuban

Revolution. The communal garden there was for him a place of relaxation, a place he went to recover from a hangover after too many mojitos or a place to chat with neighbors.

From amaranth to snow peas—vegetables, rare cultivated plants, and herbs are grown in a variety of mobile beds. Sage, Good-King-Henry, and other garden delicacies sway in tubs; squash tendrils weave around the crates. Since 2011, vegetables grow in stacked crates on terra preta substrates produced from kitchen wastes collected during the Über Lebenskunst Festival (Art of Survival) at the Haus der Kulturen der Welt (House of Cultures of the World). No longer are crops growing on purchased organic substrates but rather on the best homemade soils to be found in the area. A year later, a further source of biochar was found in using bokashi from vegetable and fruit wastes from the local food bank.

Interestingly, these green guerrillas consider the fact they are not trained horticulturists to be an advantage: "We have to be more open and seek input and advice from all different sources." As well as encouraging ecological diversity, they also promote broader social diversity. The exchange between different cultures, classes, and generations seems to occur quite naturally here.

Young mixes with old, Turks and Russians mix with Germans, and professors mix with the unemployed. Even the herbs growing in sawn-off milk cartons and sold for 3.50 euros are portable. "Mojito mint" jokes the Green Che, recalling memories from his time in Cuba.

The latest reports from winter 2015: Prinzessinnengarten will remain, in all likelihood, in its entirety.

PROTECTING SPECIES AND THE CLIMATE

THE PROTECTION OF species and the climate should go hand in hand. If you planted a 100-meter-long (328-foot-long) strip of trees on every hectare of agricultural land in Europe, which corresponds to only 3 percent of the arable land available, you would not only improve the biodiversity all over the continent, you would also sequester 20 percent of Europe's annual carbon emission in the trees, roots, and humus. A similar result could be attained if all European farmers shifted to zero-tillage cultivation and the use of organic fertilizers.

The U.S. ecologist and Alternative Nobel Prize winner Albert Bates calculated in his book *The Biochar Solution: Carbon Farming and Climate Change* that if every person on earth were to plant one tree a day, then after only two years, so much CO_2 would have been extracted from the atmosphere that a new ice age would begin. Although this is a mathematical equation rather than a practical climate change mitigation strategy, it demonstrates that with sufficient political will to reverse climate change, there are ways to do it naturally. It is possible that with carbon-smart agriculture and increased urban gardening, Europe could become almost carbon neutral, while simultaneously booking agricultural yields in many cases. We could stem the losses in biodiversity and thus rebalance our environment. Instead of naked or monocultural soils, colorful fluctuating fields with networks of hedges, field margins, and strips of trees would encourage biodiversity in a phenomenal way in the countryside. It could be so easy to save the world; we just need to give nature a chance to help us.

Every new tree, every new bush, every hedge helps nature in some way. Everywhere, in cities, villages, parking lots, lay-bys, footpaths, crossroads, playgrounds, front gardens, balconies, backyards, and roofs, there are public and private spaces to plant new trees or bushes. Every planting of a tree or hedge could be viewed as a small act of resistance against the alienation brought on by the doctrines of capitalism and the incessant drive for economic growth.

All manner of biota provide innumerable ecosystem services. Trees filter the air and water, they regulate the air's humidity, moderate the summer heat, collect dust, absorb CO_2 from the atmosphere, and offer a refuge for numerous insects, birds, and regulating microorganisms. Plants carry out a myriad of services for the benefit of all of us, which should be reason enough for all of us to nurture nature. Yet many continue to insist on economic payback on any investment, and most prefer paybacks with the shortest term and the highest return. Viewed through this capitalist lens, investing in nature often loses out to other more financially attractive options.

The village of Victor, New York, like an increasing number of other small towns across the U.S., has its very own tree board. The goal of the tree board is to develop and implement an urban forestry plan with a diverse range of regionally appropriate trees, with an emphasis on diversity and preservation of distinctive old trees. Tree boards are giving a budget of at least two dollars per capita per year to maintain and improve urban forests. If more communities across the globe followed this tree preservation and promotion strategy, not only would it improve urban landscapes and reduce the heat island effect, but the carbon sink potential of cities could be greatly improved.

Frequently nowadays, biodiversity is higher in cities than in the countryside, where monoculture farming is practiced and crops are sprayed with pesticides. Urban beekeeping on balconies, roofs, and windowsills not only produces honey but also helps prevent bee colony collapse by keeping bees in pesticide-free zones.

PROTECTION AGAINST POLLUTANTS

THERE IS A further pragmatic argument in favor of using terra preta soils in urban gardens, as they reduce the spread of pollutants. According to studies by the Institute of Ecology at the Technical University Berlin (TU Berlin), trees and shrubs are an effective barrier between busy roads and vegetable gardens.

Vegetables grown in urban areas, or near busy roads or highways, often absorb heavy metals, such as lead or cadmium, which come from paints, waste incineration, and the production of fertilizers, or polycyclic aromatic hydrocarbons (PAHs), which are found in exhaust, tires, soot, and asphalt. "Vegetables from inner-city gardens can have many times higher concentrations of heavy metals than standard supermarket products," says Ina Säumel from the TU Berlin, under whose leadership samples were taken from 28 garden sites in Berlin. The readings vary according to type of vegetable, but the location was even more decisive for the results.

However, the ecologist emphatically warns against fear-mongering as it relates to urban gardening. Risk assessment should be done as a whole; all of the positive aspects of urban gardening, such as shorter transport distances for food, CO_2 savings, movement in fresh air, communal experiences of nature, and the joys of gardening, should be considered.

Research has shown that biochar is able to tightly bind various metals, rendering them unavailable to plants. Thus, using biochar offers potential relief to urban gardeners from certain organic and inorganic toxins. Washington State University Extension researcher Linda Chalker-Scott published "Biochar: A Home Gardener's Primer," which states not only that biochar "binds and/or detoxifies heavy metals such as lead, mercury and chromium" but that it also "binds and sequesters organic contaminants such as PAHs."

Additionally, urban gardeners can safeguard against pollutants, for instance, by planting hedges or building a garden shed. They also shouldn't plant certain crops, such as lettuce, next to old walls, as these may have been painted with lead paints. All harvested vegetables should be washed, even if they are organic.

SQUARE-FOOT GARDENING, OR GARDENING ON THE GO

SQUARE-FOOT GARDENING WITH terra preta is an excellent way of growing vegetables locally, and it doesn't matter whether you do it in an urban setting or in the countryside. Mel Bartholomew, an engineer from Utah, noticed that every spring, hobby gardeners plant far too many seeds on garden beds that are overly large. Another problem he noticed is that certain crops are grown in large numbers and are ripe for picking all at once, but often much of the produce goes to seed, since it is impossible consume it all before it goes off. Often, this kind of situation results in gardens gone awry and gardeners who lose the desire and energy to carry on gardening.

Bartholomew's answer was to develop what he dubbed "square-foot gardening," so called because the growing area is divided into small square sections, typically a foot long on every side. Square-foot gardening merges old ideas with new ones, resulting in simple, successful, and pleasing mini-gardens. Perhaps the most important aspect of square-foot gardening is psychological—enabling city dwellers to use a small defined space with a clear plan of how to make use of it intensively and grow at least some of their own food. Carbon-rich terra preta soils are the right choice to make square-foot gardens not only highly productive but also more fun!

One square foot is maybe a little too small to produce significant amounts of food, but 1.2 by 1.2 meters (4 by 4 feet) will produce a substantial amount. It could also be longer but not wider so that you can easily reach everywhere from the sides, say 2.4 by 1.2 meters (8 by 4 feet). Borders can be made from wood, roof tiles, or stones, and filled to a depth of 20 centimeters (8 inches) with homemade or purchased terra preta. Kick-starting with 2 to 4 liters (0.5 to 1 gallon) of diluted urine may be a good idea.

Square grids made of wooden slats measuring 30 by 30 centimeters (1 by 1 foot) are placed on top of the gardening bed. Small plants and seeds are placed in these minisquares. This grid structure is an enormous help as a guide to what is actually growing, what can be picked, and what needs to be replanted. People who cannot easily bend down can elevate the beds with a table.

Each square can be used for a variety of different vegetables suited to the gardener's taste. Each square could contain 4 kohlrabies, 1 broccoli, 4 chards, 1 cauliflower, 4 celeries, 16 radishes, 9 bush beans, 16 leeks, 16 onions, 9 spinach, or 1 potato plant. On shadowy north-facing sites, you can plant climbing plants, 8 runner beans, 1 squash with climbing frame, 8 pea plants, 1 corn, and so on.

In this way, it is possible to grow 20 different plants on one 3-meter-square (10-foot-square) bed. If one crop happens to fail, it is easy to replant something else immediately. A square garden can provide the basic needs of one person, who can pick herbs every day, lettuce three times a week, and other vegetables twice a week during the growing season.

Squares that have been harvested can be revived with terra preta substrates, bokashi, or compost and, depending on the season, immediately replanted. With a little planning, gardeners can use crop rotation and mixed cultivation and harvest their bounty three times a year (maybe radishes, then bush beans, afterwards brussels sprouts, or first spinach, followed by celery, and finally corn.) With a bit of luck and good weather, in some temperate climate zones it is even possible to have four harvests.

The location of the square garden should be as close as possible to the kitchen or front door—as you will more easily notice if something needs to be picked, weeded, or replanted. The diversity of possible plants that are able to thrive in square-foot gardens allows gardeners to learn a great deal.

AGROFORESTRY IN NEPAL

MANY RURAL AREAS in Nepal are suffering from the exodus of laborers to urban areas, leaving large swaths of farmland barren. The previously tended terraces are now at high risk of erosion during the rainy season. In early 2015, 40 farmer families joined to plant 10,000 trees, with the intention of building a robust agroforestry community that will help both the environment and the local economy.

A variety of trees were planted, including cinnamon, mulberry, paulownia, melia azedarach, and magnolia champaca. Trees were selected based on the products they provide to locals, as well as how well they interact with other tree species and other companion plants. Products derived from this reforestation project include essential oils, leaves for silkworms, perfume, honey, timber, and livestock fodder. These trees also provide critical ecosystem services such as erosion control, improved water management, biodiversity, humus creation, and reduction of landslide risk.

Each tree was planted in a small pit filled with urine-enriched biochar, which was made from a local invasive species known as eupatorium, or more commonly called "forest killer." All biochar was produced in Kon-Tiki kilns and quenched with cattle urine. Roughly 4.5 tons of biochar was used, which sequestered the equivalent of 11 tons of CO_2. Using conservative estimates for each variety of tree planted, each year approximately 158 tons of CO_2 will be sequestered in these agroforests, the equivalent of taking 50 cars off the road! Impressively, the survival rate for these trees was more than 95 percent in the first three months, a large improvement over the average survival rate of 50 percent found in other reforestation projects in Nepal over the past 20 years.

This agroforestry project brings multiple benefits, including climate protections, job creation, and vastly improved ecosystem services. Sustainably focused companies looking to offset their greenhouse gas emissions, especially those that rely on the various products produced in agroforestry environments, are beginning to fund these types of environmental revitalization projects. There are millions of hectares just waiting to copy the concept.

AGROFORESTRY CULTIVATION

FORESTS ARE CRUCIAL to earth's climate and to humanity's survival. They store vast amounts of carbon; provide nutrients, energy, building materials, and habitats; stabilize weather; and offset temperature fluctuations. The ancient inventors of terra preta certainly didn't understand all of these relationships, but they must have realized that trees can safeguard the success of their field crops. Banana, mango, and avocado plants flourished under the leafy canopy of the jungle giants.

Agroforestry cultivation is a "multistory" system of cultivation involving growing trees in combination with food crops or pastures. It is an ancient growing technique that is enjoying a resurgence in various places around the world. In combination with terra preta, it provides multiple benefits related to protecting and improving the environment, while improving farmer livelihoods. At the moment, however, agroforestry practices are not widespread and there are many challenges to increasing participation around the world. Markets for niche products need to be developed; successful regional agroforestry examples need to be tested, tweaked, and rolled out; and farmers must be educated about what combinations work in their environments, how to care for the various trees and plants, and what post processing might be required.

There are many positive signs that agroforestry is catching on. According to satellite data released by the Food and Agriculture Organization of the United Nations (FAO), trees in fields are making a comeback, especially in Latin America,

Africa, and Southeast Asia. According to a report by the scientific journalist Volkart Wildermuth, farmers in Brazil have rediscovered that banana leaves can be a valuable source of fertilizer. In Africa, many thousands of women earn a living by converting the nuts from karite trees into shea butter. In Burkina Faso, Mali, and Niger, the acacia, *Faidherbia albida*, is used as a fertilizer tree—crops grow more profusely around the base of this exceptional legume that has the unusual ability to store nitrogen in its roots. Additionally, this tree sheds its leaves during the rainy season, when other trees are just beginning to turn green, whereas during periods of drought it supplies shade and nitrogen to crop plants as well as fruit for grazing animals. The area surrounding these tree has visibly more humus. In some areas of the Sahel region where farmers are making use of agroforestry, desertification is slowing and in some cases being reversed.

In southern Africa, new agricultural methods have evolved around the acacia tree: conservation farming. The key elements are cultivation of acacia, low-impact soil tillage, and selective crop rotation. In the mid-1990s the Conservation Farming Unit began by training 12 farmers, who because their yields increased, enthusiastically passed on the knowledge to others. The method snowballed, and since 1996, hundreds of thousands of farmers have adopted this method and passed it on.

Dirk Freese, professor of soil conservation and recultivation at the Brandenburg University of Technology in Cottbus, Germany—a region wrecked by the ecological consequences of coal mining—is hoping for a similar snowball effect. The remaining slag heaps and pits are biologically dead, the soil unable to foster plant growth, much of it already eroded by the wind. Freese tested whether implementing agroforestry structures could transform soils low in nutrients into high-nutrient soils and stop erosion.

Young poplar and robinia trees were planted in 12-meter-wide (39-foot-wide) strips a few hundred meters long along one part of the roughly 2,000 hectares (5,000 acres) of terrain, which is run by a farmers' cooperative. In between these rows, lupines and alfalfa, classic nitrogen-fixing legumes have been planted. This strategy seems to have reduced erosion caused by wind, improved rainwater retention, and increased humus content within a few short years. Every two to three years, the cooperative harvests the trees, which they convert into fuel.

In France and Great Britain, too, many groups have been working successfully with agroforestry for more than twenty years. "Trees constantly feed the soil throughout their whole lives," explains Christian Dupraz, research director of the French National Institute for Agricultural Research (INRA) in Nils Aguilar's documentary film *Voices of Transition*. "Several times a year, a tree loses its feeder-roots, which enrich the soil with organic matter. This process also aerates the soil, allowing organisms to breathe." A field planted with cereal grains with a scattering of trees is 60 percent more effective than a conventional monoculture field. Fewer pesticides are needed, as the trees serve to attract bats that eat insects and other pests. Trees "can influence the microclimate. Extreme temperatures are reduced and hot and drying winds are stopped. Trees help to regulate weather and cycle water. They also improve global climate by capturing carbon."

For Martin Crawford, director of the community Agro-forestry Research Trust in the British town of Totnes, agroforestry is "the most efficient agricultural system there is," because plants mutually protect and fertilize.

TREES AS RAINMAKERS IN THE SAVANNAS OF COLOMBIA

SOME 30 YEARS ago, the Orinoco savanna in eastern Colombia lay bleak and desolate under the horizon. Settlers from Europe destroyed the rain forests to make way for cattle farming, driving off the indigenous communities in the process. Because of over-grazing and no longer having the protection of trees, the fertile topsoil was soon washed or blown away. The soil became impoverished and exposed to the equatorial sun. What was once lush rain forest and the habitat of countless flora and fauna could no longer sustain much life. Fortunately, today rain forest grows there once again, and with it biodiversity and rain have returned. The forest now supports more than 200 farmer families, who on top of growing fine products from the forest gardens also sell drinking water to Bogotá.

In 1984, when the Colombian Paolo Lugari came to the village of Gaviotas with the vision of restoring the land that had become so depleted, most experts considered it impossible. The soil was too acidic, the sun too hot, and water too scarce. A mere 10 years later, 5,000 hectares (12,000 acres) had already been reforested; to date, the total has risen to 8,000 reforested hectares (20,000 acres). The planting of trees brought back the rain forest with all its diversity astonishingly quickly. The soils have regenerated and once again provide clean water. The government has already handed over a further 45,000 hectares (100,000 acres) to be reclaimed by rain forest flora and fauna. An area of 6.3 million hectares (15 million acres) of surrounding land could follow and also become forest gardens.

Lugari explains to foreign visitors in simple words why the removal of the rain forest led to desertification: "Once the protective blanket of vegetation vanishes, the tropical sun heats up the ground. When the ground temperature is higher than air temperature, rain evaporates without having time to be stored in the ground, meaning that there is too little water available for plants. When there is rainfall in the desolate savanna, it is short and incredibly heavy, which can lead to disastrous compaction or erosion on bare ground, and desertification advances."

Lugari chose the Caribbean pine as the pioneer for the replanting program. The taproots of these trees can absorb water from deeper levels of soil and at the same time improve the storage capacity of the soil during the infrequent but heavy rainfall. Additionally, the delicate needles of the Caribbean pine evaporate relatively little water. Before the trees were planted in the depleted and acidic soil, the roots were treated with mycorrhiza. As soon as the saplings were large enough to offer some shade, the surroundings became cooler and moister. Seeds that had lain dormant in the soil or transported there by birds could germinate. Step by step, the rain forest with all its majesty returned. Today in Gaviotas, there are again no less than 250 different species of trees.

Thanks to the evaporation above the trees and the lower temperature of the soil, the atmosphere above the forest is cooler than in the surrounding savanna. Perhaps most importantly, water evaporation in the forest creates new currents of air mass. Hot air, driven by savanna winds, rises over the forest and cools down, leading clouds to form, which bring localized rain events. The amount of rain has increased by 10 percent in only 10 years; maybe even more importantly, the rain falls more regularly throughout the year, instead of only in devastating torrents.

Dried up river sources began to flow again. Crystal-clear water from the region is now sold as drinking water in Bogotá. The resin of the Caribbean pine serves as a raw material in dye and paper industries. Oil from palms is processed into fuel that now no longer has to be transported in from great distances. Within 20 years, the soil has become so fertile that land prices are 3,000 times higher than the surrounding savanna.

The inhabitants of Gaviotas continue to follow the traditional knowledge of the indigenous people. Forest gardens are planted using heritage species in mixed cultivation. They use terra preta in the vegetable beds and produce their own biochar from the large quantities of residual wood available from annual pruning. More than 200 families in Gaviotas are almost exclusively sustained by locally grown forestry byproducts.

Reforestation is also an important topic at Colombia's San Gil University. The small town, 400 kilometers (250 miles) northeast of Bogotá, is situated in a coffee-growing region that is also attractive for tourism. The university develops regional initiatives on the notion of a solidarity-based economy and in recent years has had much success with women's cooperative projects; another new forest now prospers in Barichara, not far from San Gil, thanks to one of these projects. The women used the knowledge of the local population in the selection of tree species, set up water basins typical for the region, and encouraged schoolchildren to become involved in the planting and care of trees. The experiences at Gaviotas and Barichara were mutually beneficial and have served to inspire similar projects around the country.

There are still many forest gardens in the vicinity of San Gil. Coffee is grown mostly in smallholdings beneath shade tress, with papaya, citrus fruits, manioc, corn, and beans. Some 10 kilometers

(6 miles) outside San Gil, Isabel and Thomas Garcia produce coffee and cocoa on their 30-hectare (75-acre) organic farm, without the need for pesticides, thanks to the biodiversity of their forest garden. For many years, they convert organic wastes into bokashi, which they then used as fertilizer for their coffee plants, and they have been producing terra preta since 2012.

CHAPTER 6

Old and New Sanitary Systems

Terra preta enables you to upcycle your own excrement, thereby reconnecting nutrient cycles. Homo, humus, and humanitas—it is no accident that the three words share the same origin.

IN THE AZTEC language Nahuatl, the word for gold is "*teocuitlatl*," which means "the excrement of gods" (or "holy shit"). For the Aztecs, gold had a very different meaning than what it meant for the conquistadors, which tragically led to many fatal misunderstandings. To indigenous people, gold referred to the excrement of their sun god. The shiny metal that reflected the sun symbolized for them the eternity of the divine cycle. Gold, silver, and precious stones were abundant in the kingdom of the last ruler, Montezuma. The Aztecs obtained most of their "holy shit" from rivers and used it for making statues and ornaments. In their eyes, this type of gold had no material value, as their trading currency was cocoa beans. At that time, a small piece of woven white cotton was valued more highly in cocoa beans than a small gold statue.

In 1519, Montezuma greeted the Spanish conquistador Hernán Cortés and his 500 soldiers with gifts of gold, which grievously led to the beginning of the end for his people. The incessant greed of the European colonizers was immediately aroused, and not long after, they were transporting shiploads of stolen melted-down treasures of the defeated Incas, Mayas, and Aztecs to Europe. Colonization advanced throughout the continent, leaving a trail of looting, exhausted soils, and misery. In the process, the conquistadors completely missed the meaning of the true El Dorado—the capacity of many indigenous civilizations to manufacture black gold, exceptionally fertile soils, from little more than their own excrement, other organic wastes, and biochar.

The misbegotten quest for El Dorado, this chasing of shiny materials with little intrinsic value and use for humans beyond the mere fact of its scarcity, continues to echo in our current consumer society. On the one hand, the centuries-old quest for *teocuitlatl*, or so called gold, aroused greed that continues to this day, spawning an endless series of speculative derivatives that should be flushed down the toilet. On the other hand, the real excrement of humans is considered worthless, though it is capable of nourishing all manner of flora and fauna.

The aversion to feces is not completely unfounded—human feces can cause parasitic infestations, diseases, and plagues. For this reason, feces should never be used directly as fertilizer. If, however, feces is fermented with biochar, sanitized, and buried in compost for at least a few months, it becomes part of the black earth that smells as sweet as a forest floor.

DARK MATTER—A SHORT HISTORY
OF THE CALLS OF NATURE

NOT ONLY THE Mayas but also many other indigenous cultures discovered, or at least suspected, the secret of turning human excrement into fertile soil. In ancient China, Japan, and Korea, urine and feces were carefully collected; in China today, they still refer to urine as "fertilizing water." The Romans gave the god Saturn the epithet Sterculius or Sterces, from "*stercus,*" meaning manure, and he taught people how to fertilize their fields with feces.

In ancient Greece and Rome, there are also reports of urine and feces being saved in portable vases, buckets, or other such vessels. Ancient stone toilet seats consisted not only of round holes but had deep, forward-sloping, rounded notches. Most likely these served as slots for placing pottery vessels used to collect urine. Roman emperor Vespasian, who reigned from 69 to 79 AD, was quoted as saying, "*Pecunia non olet*"—"Money doesn't stink"—which is attributed to the fact that ancient Roman collection vessels for urine were found in public conveniences. Urine was considered a valuable raw material for agriculture and was even transformed into a cleaning agent and was therefore taxed.

Although the first flushing toilets appeared in Europe in the sixteenth century, it took 200 years for them to become commonplace. Even at the beginning of the twentieth century, in both large and small cities, the products of human metabolism were carted off and used as much-sought-after fertilizer. Collection systems were organized much in the same way as today's garbage disposal and had the advantage

DRY TOILETS INSTEAD OF CONTAMINATED CROPS IN MEXICO

MEXICAN ARCHITECT CÉSAR Añorve has been concerned for a number of years with the contamination of water by the centralized sewer systems. The quiet, somewhat shy architect and inventor lives with his family in a house disconnected from public water with a large green plot of land in Cuernavaca, south of Mexico City. He has been busy working on practical alternatives ever since the region became a horrific example of water contamination.

The sewer systems of Mexico City are prime examples of poor central planning. Greater Mexico City has a population of more than 21 million and climbing. Every day brings more newcomers who build unauthorized houses. As a result, the centralized sewage system is hopelessly overburdened by the collection of excrement. Flushed with already scarce water supplies, human sewage is pumped up into the mountains surrounding the city, where the wastewater is cleaned and then pumped back down to the city again. All of this comes with drastic consequences. The water quality of the capital is so contaminated that anyone who can afford it buys drinking water in plastic bottles. A large proportion of the untreated sewage flows unchecked into the once-fertile plains beyond Cuernavaca and over the years has contaminated both the groundwater and the soil. Here they are permitted to grow flowers but not food, yet the situation is made worse by the use of synthetic fertilizers and pesticides.

César Añorve is an out-of-the-box thinker and tinkerer. He installed simple self-sufficient rainwater collection systems and water pumps driven by pedal power. By using water vortices, he was able to purify the rain and shower water.

He shows off his dry toilets that he built himself, which enable urine and feces to be separated. Asked whether it was more difficult for women to use his design than men, his daughter-in-law replied, with much animation, that it was just a question of practice. If a woman really wanted to, she could quickly learn to use the appropriate outlets for urine and feces.

Using simple means and customizing according to client's wishes, Añorve has designed and built a wide variety of different toilets. The molds he uses are made from plastic reinforced with fiberglass into which he pours concrete. He can also produce plastic or ceramic models. He sells upwards of 200 dry toilets per year, from which he is able to make a modest living. The dry matter is covered with charcoal powder to sanitize it and contain odors while it ferments. Añorve intends to make the production so simple that dry toilets could easily be produced in small regional workshops anywhere so that the use could spread to areas where money is even scarcer than water.

A large-scale shift to such ecologically sound sanitary systems will not be all that easy, since Mexicans, like many people worldwide, prefer the convenience of flushing their excretion away with drinking water. Despite thoroughly well-thought-out dry toilet designs, there are still problems with targeting, incorrect use, and inadequate maintenance. Dry toilets work only when the users are conscious of and have respect for the toilets' main advantage, which is making use of excretions to create healthy fertile soils on which to grow healthy food.

The bedding for human excrement plays a critical role. With charcoal powder and microorganisms, comparable to effective microorganisms, César Añorve managed to ferment human wastes in sealed vessels, which were well suited for processing into terra preta. In his garden, there is a huge clay pot with a capacity of 300 liters (80 gallons). In this pot, he shovels in layers of foliage, garden soil, charcoal powder, and his humanure bokashi. He compresses the mix for further fermenting and maturing and seals it. There are no unpleasant smells; only the familiar acidic smell of bokashi is perceptible.

Such sanitary systems could solve many problems simultaneously in squatter settlements, slums, and refugee camps. When human excretions are fermented in combination with charcoal powder and in the process made sanitary, odors and transmittable diseases can be prevented and precious water saved. Crops can be fertilized, and families can earn a living with less exposure to harmful fertilizers and pesticides. And it all helps in the fight against climate change.

over cesspits of not polluting groundwater. In Leipzig, Kiel, and Berlin, they adopted the Paris model and transported the contents of the buckets to composting facilities, where they were processed with other aggregates and turned into fertilizer. Outhouses were commonly located in places where liquids and solids could be separated with at least a modicum of success. On farms, they used dung piles where urine could drain off into cesspits while solids remained on the manure heaps.

Chamber pots were used for collecting urine, and outhouses were generally for solids. There were even urine receptacles that women used in public beneath their clothes. A *bourdaloue* in the form of a gravy boat was fashionable for a short time in the eighteenth century. It was discretely placed in a leather cover and could be used as a traveling toilet. The urine that women collected in these devices was probably not emptied into the outhouse but disposed of directly in a garden.

Albrecht Thaer, who is considered the founder of agricultural science, wrote in 1806: "With very few exceptions among the people of Europe, there is a curious and highly detrimental prejudice, which prevails against the use of human excrement as fertilizer. If it were carefully stored, mixed with vegetable matter, and applied to fields at just the right moment of fermentation, all other fertilizers would be rendered unnecessary." In the nineteenth century, there were fierce disputes about whether it was the duty of a growing city to produce fertilizers for agriculture, but ultimately, the evolving mineral fertilizer industry prevailed.

With the spread of central drinking water supplies towards the end of the nineteenth century, many of the epidemic problems of the time were solved; however, at the same time a new problem was created, namely wastewater. The old waste collection systems could no longer cope with the massively increased amounts of wastewater. Chamber pots were considered old-fashioned; dry toilets lapsed into oblivion. A similar fate awaited outhouses, which were still to be found at the end of every patch of garden at the beginning of the twentieth century, with the goal of producing fertilizer for the garden.

Water closets heralded the advent of capital-intensive waterborne sewage systems and eventually large wastewater treatment plants. Some 80 percent of the capital investment for our wastewater systems has been buried in the ground in the form of pipes. Treatment plants nowadays consume enormous amounts of energy to extract nitrogen, carbon, and phosphorus from wastewater. The resulting sludge is then, in many cases, dried and incinerated at great cost, because it is contaminated with pollutants and can no longer be used as fertilizer.

The remaining 95 percent of purified water is channeled to the sea via rivers carrying trace nutrients, nondegradable pharmaceuticals, and other materials along with it that the often outdated wastewater treatment plants weren't able to capture. During heavy rainfall, diluted wastewater flows untreated into the waterways, polluting everything in its path. We have become used to this technology that wastes both drinking water and fertilizer. Each year, we pay a great deal of money to manage wastewater in the misguided belief that this is better for the environment. The wastewater treatment models of the twentieth century are obsolete and should be updated. If we didn't mix urine and feces with other wastewater from the kitchen and bathroom, household wastewater would be much simpler to process and reuse. Eighty percent of the nitrogen found in wastewater comes from urine; in a world with a finite supply of fossil fuels that are needed to produce synthetic nitrogen, these liquids should not be considered waste.

In our society, the taboo surrounding excrement has led to a taboo of discussing toilet-related topics. We no longer talk about our daily relief and scarcely even think about it. Many developing countries are keen to copy this sanitary "advance" and build centralized sanitary systems, even if they possess neither adequate water supplies nor the electricity necessary to drive the pumps and treatment plants. In addition to being costly, in some cases they even do more harm than good to the environment, especially when equipment fails or employees are improperly trained.

Depending on drinking and eating habits, a person will excrete between 0.5 and 2.5 liters (17 to 85 fluid ounces) of urine and between 60 and 250 grams (2 to 9 ounces) of feces daily. That means that every year, every human produces roughly one 60-liter (15-gallon) barrel of solids and nine 60-liter barrels of liquids, all in all ten full barrels. With the nutrients from these barrels, we could fertilize 100 to 1,000 square meters (1,000 to 10,000 square feet) of soil annually, an area large enough to grow enough food to sustain us.

Sadly, urine and feces are considered to be evil-smelling problems that must be banished. We are led to mix them with perfectly good drinking water, instead of redirecting them as important raw materials back to the natural cycle.

Essentially, we are creating human sludge. Nitrogen and water in urine quickly cause feces to rot, which in turn encourages the proliferation of pathogens.

Humans need to be acutely aware that every liter (0.25 gallon) of human waste that lands in the waterborne sewage systems is diluted with 80 to 100 liters (21 to 26 gallons) of drinking water. In many developed countries, the water used to flush toilets is, generally, of drinking quality. Drought-prone countries such as Australia have learned to understand the precious nature of clean water and using it to flush waste is not high on the list of priorities. Conserving water and harvesting nutrients from human waste must become part of the new millennium thinking if we are to survive and thrive on an increasingly parched planet.

THE FEASIBILITY OF ALTERNATIVE TOILETS

IN COUNTRIES WITH low population density, such as Scandinavia, scientific experts have been converting old outhouses into proper compost toilets. Today, there is a whole series of aesthetically designed urinals and toilet systems on the market that function perfectly well without water. The Hamburg-based industrial designer and author Wolfgang Berger offers a whole range of these modern dry toilets. Compost toilets, with or without separating mechanisms, provide valuable raw materials for humus. One of the drawbacks of these open compost systems, however, is that they need some form of ventilation—either ducts or electric extractor fans. Energy and nutrients such as nitrogen are lost, and the likelihood of insect infestation increases. But there are also

AYUMI MATSUZAKA AND EDIBLE ART

AYUMI MATSUZAKA, WHO was born in the Japanese city of Naga-saki in 1978, has caused quite a stir with her extraordinary public art projects. Under the title *All My Cycle*, she has been producing terra preta with her own urine, fertilizing vegetables with it, eating them, and then using her own excretions again to make fertilizer.

The 37-year-old studied art in Tokyo, Venice, and Weimer. For the brief times when she's not traveling, she lives in Berlin. In her project *Time Bank* in Finland, she made jam that served as regional currency, bringing together people who lived far apart. For another project, in Vietnam, called *Dream Collector,* she invited people to talk about their dreams and subsequently embroidered images from the dreams on bed sheets. In France, along with two chefs, she displayed 100 pieces of toast decorated with different jams.

In 2010, she became interested in the art of making terra preta, which she learned from black earth pioneer Jürgen Reckin in Marienthal, north of Berlin. For an art performance, she planned to eat vegetables that had been fertilized with her own fermented urine. The first attempt failed, because the soil hadn't matured suf-ficiently, and her vegetables died. She brought her plants in crates to Prinzessinnengarten, learned more about the art of producing terra preta, and has since passed on her knowledge in various workshops. Now everything runs much more smoothly.

Ayumi Matsuzaka, cheerful and uncomplicated, would like to make us less afraid of our own excretions—Asian cultures are much more open about this subject. In New Delhi, another art project, *Microbe Intervention,* involved giving Indian families starter kits with biochar and microbes with which to produce their own terra preta on balconies and gardens. In September 2012, as part of the *Examples to Follow!—Expeditions in Aesthetics and Sustainability* in Beijing, she held a workshop on how to close natural cycles using the stacked crates system.

Her project *A Basic Exercise* ran in Stuttgart until 2013. The artist challenged visitors to bring their own excretions with them to the exhibition, and sure enough, a few dozen brave souls responded to her plea. The artist put an end to the unwanted smells with biochar powder and effective microorganisms, and participants could observe the fermentation process, receiving in return a pot with a flower grown in their own byproducts.

airtight sealable toilets, most of which are advertised as portable camping toilets. With the right material, smells can be avoided and your own inner assets refined.

As we have already said: human feces is not absolutely required to make terra preta. Still, you should give some thought to the absurdity of flushing away your own excretions, which are rich in nutrients, and in the process using up one of our most precious resources—drinking water. Even if only the urine was recovered, most of the fertilizing value would already be recycled. To collect the urine in a bucket filled with biochar does not produce any bad smell and no further composting treatment is necessary. It is very easy to do and an efficient, safe, and easy-to-apply organic fertilizer.

Using old knowledge and his own reflections on the deficiencies of modern compost toilets, Haiko Pieplow came up with the idea of developing an anaerobic dry toilet for personal use. Horticulturist and EM consultant Marko Heckel acted on the idea, and in 2009, his company TriaTerra launched an anaerobic dry toilet. The basic idea is remarkably simple, and variations could be used worldwide.

The anaerobic dry toilet is something very personal. It consists of a sealable feces container and a urine receptacle that needs no pipes, electricity, or infrastructure. It can be placed anywhere and even in summer temperatures remains odorless and hygienic. Each member of the family can have his or her receptacle if so desired, and guests have the option of using a conventional toilet.

For a terra preta toilet, as the anaerobic dry toilets are also known, you will need at least four airtight, sealable containers. The containers should have volumes of 20 to 60 liters

(5 to 15 gallons) so that they are easy to move and empty. For collecting feces, you should have two 20-liter (5-gallon) containers with wide openings that can be well sealed. Plastic buckets or earthenware pots are well suited to this purpose. For storing urine, 60-liter (15-gallon) drums have proven to be useful. Rain barrels from hardware stores or any other large industrial plastic drums can be easily converted.

Toilet paper also ends up in the solids container. Each time the dry toilet is used, the results are covered with biochar. The biochar for the solids container should be soaked in a solution with effective microorganisms to rapidly induce the fermentation process. Rock dust and/or 10 percent garden soil can be added to the biochar mix. Fermentation with kitchen wastes is also possible, but the increased volume requires more frequent changing of the solids containers. While the bucket is filling, the previously filled container is fermented to humanure bokashi, preferably in a warm place. After a few weeks, a characteristic white mycelium coating forms on the surface.

It is very important that little or no urine reaches the feces container. If urine and feces are stored together, it will generally result in methane-generating decomposition, stenches, and contamination. As with composting, unpleasant smells are a sign that something has gone wrong. What helps is again a dose of charcoal powder and EM solution.

Urine leaves the body almost germ free. Medications, which are broken down considerably better in soil than in water, are also expelled via urine. Mixed with biochar or diluted with water, urine can be used directly as fertilizer (see chapter 3, page 137).

Feces should never be used directly as fertilizer. On the one hand, there is the danger of infection, and on the other hand, valuable nutrients could be leached out by rain. Lactic acid fermentation initiates the sanitation process, but the process will only be fully complete with humification. Feces consists of undigested food remnants, dead cells, and a healthy dose of anaerobic microorganisms from our intestines. Biochar offers these microorganisms a habitat, enabling them to perform their beneficial activities. Each layer of biochar in the container of feces is successively colonized, and the addition of EM assists the process.

After one to two months, the fermented humanure bokashi can be put in the compost heap to finalize the humifying process. It should be layered in with kitchen bokashi and garden soil so that it can humify quicker. The proportion of biochar should be at least 10 to 15 percent of the total volume of the compost heap. The proportion of humanure bokashi shouldn't exceed 30 percent. Humanure and kitchen bokashi are much better food for soil organisms than the same materials raw or rotting.

In subsequent years, the humified biochar substrate matures to terra preta, provided we practice sustainable cultivation as described in chapter 4, with mixed cultures, mulch, and organic fertilizing.

HUMUS FROM HAMBURG'S MAIN STATION

THE PLACE WHERE the first modern urban wastewater recycling system originated is pretty unassuming—the public restrooms at Hamburg's main railway station. Every year, some 200,000 people leave their nutrients here, which are then sanitized, stored, and converted into terra preta. Peter-Nils Groenwall, the environmental authority official responsible for water conservation, cemeteries, public parks, and small garden plots are fertilized.

When Groenwall visits the facility that he conceived, the toilet attendants very kindly welcome him. The employees of the public restrooms in Hamburg are very grateful to the socially minded environmental expert. The economist and sociologist was responsible not only for preserving their jobs but also raising their prestige. Increasingly, scientists and journalists come to observe and learn about the transparent wastewater pipes of Hamburg's netherworld.

Everything is astonishingly clean to assure the biochemical processes of the waste treatment. Men urinate in no-flush urinals so that liquids and solids can be separated and collected in storage tanks. Separation is not possible, however, for women, but the "green gain" system, which allows a choice between a short flush for liquids and a longer flush for solids, reduces the amount of water flushed to 2.5 liters (0.5 gallon) per flush. Additionally, microbiological filters in the cellar almost completely separate the waste flows.

Once the feces arrive in the tanks below the restrooms, it is automatically mixed with biochar and fermented with effective microorganisms. Accompanying research by Ralf Otterpohl from

Hamburg's Technical University and Monika Krüger from Leipzig University found that biochar and EMs ensure a reliable, odorless sanitization and neutralize harmful germs. Afterwards, the solids are taken to composting units where the human wastes are processed into humus. Groenwall's units save not only drinking water but also a great deal of materials. The pipes are thinner and as there are no deposits building up, there are hardly any maintenance costs.

Since the committed environmentalist assumed responsibility for the almost 200 public restrooms in the 1990s, which he claimed were filthy then, he has had them all converted to the "green gain" system. This includes toilets on the elegant Jungfernstieg, the foremost boulevard in the city, and at Hamburg's bathing lakes, where restrooms are now run by solar-driven pumps. Money spent on water, electricity, pipes, and waste disposal has been reduced from 3 million euros to 631,000 euros, a reduction of nearly 80 percent!

Interestingly, because the new system is cleaner and leaves a more sanitary impression than the old one, maintenance and repair costs to toilets related to vandalism have also declined. In the long term, Groenwall would like to turn all 200 municipal sanitary units into humus producers, developing a new urban circular economy. Groenwall never speaks of feces but rather of human nutrients, a concept we might all consider embracing.

A Handful of Black Earth as Symbol

T HE CLIMATE CATASTROPHE seems at the moment to be inevitable. Big Ag is causing increasing desolation to farmers, especially subsistence farmers. Global CO_2 emissions are growing, as are famines. The dire warnings of international research groups about the collapse of ecosystems are becoming louder.

Some of them are already saying that at some stage we have to embrace climate engineering to reverse global warming. By climate engineering, they generally mean large-scale industrial geochemical intrusions into the earth's cycles, such as injecting sulfur dioxide gases or spraying nanoparticles of aluminum into the stratosphere to reflect solar radiation or "fertilizing" the sea with iron or phosphorus to stimulate algae blooms and thus withdraw CO_2 from the atmosphere. All these suggestions have enormous ecological and social risks, as the effects are unpredictable and irreparable damage could be caused to the ecosystem.

The ancient terra preta technique is one of the few exceptions to this series of high-tech, high-risk rescue scenarios. Terra preta, or biochar and planting of trees, could be called ecological engineering, as it works with nature, not against it, and generates a veritable cascade of beneficial side effects. It brings carbon dioxide in the form of carbon back to the place where it can be stored—beneath the ground. It makes soils fertile again and encourages microbial diversity, improves water management, sanitizes problematic materials, provides us with healthy food, and frees farmers from dependence on the dubious and expensive products of Big Ag. We have already described in chapter 3 how it can crucially lessen, if not stop, the effects of global warming.

James Lovelock, the proposer of the Gaia hypothesis, is of the same opinion. In an interview published in *New Scientist* in 2009, he stated: "There is one way we could save ourselves and that is through the massive burial of charcoal. It would mean farmers turning all their agricultural waste—which contains carbon that the plants have spent the summer sequestering—into non-biodegradable charcoal, and burying it in the soil. Then you can start shifting really hefty quantities of carbon out of the system and pull the CO_2 down quite fast."

Humanity cannot solely rely on help from politicians to reverse course on climate change, as there are increasing signs that national and international institutions approach the climate problems in the wrong way and counterproductively. Abstract calculations of CO_2 rarely help and don't address climatic damage that we are already experiencing. Fortunately, scientists update methodologies and calculations that previously have failed to include emission sources that were misunderstood or unknown. The use of corn to produce ethanol is a classic example. On paper, CO_2 emissions are saved compared to using fossil fuels—at least in some scenarios—meanwhile, food is being burned to produce energy while the soil is being exhausted and eroded. As previous cultures have already learned at their peril, without fertile soil, humanity cannot survive. While we focus on averting climate chaos, we must simultaneously ensure that our life-producing soils are not sacrificed.

The soil stores more carbon than all the oceans and forests combined. Much more could be stored beneath our feet, but the need to feed, clothe, and house an ever-increasing population seems to prevent it. Paving over forests prevents it. Using agricultural methods that may boost yields in the short term but exhaust soils in the long run prevents it. Every day, in every country, untold thousands of acres that could store carbon fall prey to building projects for roads, airports, industrial complexes, and housing estates. The land that remains is mostly tended with the dubious help of agro-industries and hybrid monocultures. Areas that are transformed into monocultures can no longer "breathe," negatively impacting regional climate and water cycles, and the ecosystems are severely disrupted.

If the world's greatest climate transgressors were able to unite and rigorously support financing for agroforestry systems, organic cultivation, and terra preta techniques to stop soil degradation, the world would be halfway to being saved from climate chaos. If they were to decide to introduce pollution taxes, to stop direct and indirect subsidies to fossil energy, and to ban financial speculation with land and food, global salvation would be at hand.

Organic farmers know how to retain and promote the diversity of soil life and increase humus content and in the process provide enough healthy food to feed a growing global population. "Soil quality, the productivity of the soil—that is the most important topic of all for the survival of the human race," confirms Ralf Otterpohl from Hamburg's Technical University.

Each and every person using terra preta techniques has the ability to close nutrient cycles and avoid the use of finite or high–carbon footprint resources, thereby making their own heartfelt contribution to mitigate the damage humans have caused to the planet. As a bonus, gardening with black earth brings joy, reinforces independence and self-sufficiency, opens up new insights into the miracles of nature, and fosters well-being and health. And sharing your newfound knowledge of terra preta with others is fun and enhances community.

A handful of black earth could become the sign by which we recognize one another. Let's try it!

ACKNOWLEDGMENTS

WE WOULD LIKE to thank the following well-known and not-so-well-known helpers without whom not only this book but also human life itself could never have materialized. Among the most noteworthy are:

Lumbricus rubellus
Lumbricus castaneus
Lumbricus terrestris
Lumbricus badensis
Lactobacillus casei
Lactobacillus plantarum
Saccharomyces cerevisiae
Chloroflexacea
Chlorobiacea
Chromatiaceae
Chloracidobacterium
Heliobacterium
Rhodopseudomonas palustris

(Earthworms, lactic acid bacteria, yeasts, and green and purple photosynthesis bacteria.)

INDEX

absorption capacity, 94

acacia tree *(Faidherbia albida)*, 170

Agnihotra bowls, 80

agriculture: community gardening, 68; conservation farming, 170; forest gardens, 10, 39, 43, 44, 175–76; natural farming, 98, 102, 103, 139; organic farming, 26, 28, 71, 125, 202; permaculture, 151–52, 153; small-scale, 25–26; traditional methods of, 12–13, 25–26, 29–30; window boxes, 156–57. *See also* Amazon cities; Big Ag (agro-industrial complex); climate gardening; monoculture farming; urban gardening

Agriculture at a Crossroads (IAASTD), 25–26

agrochemistry, xvi

agroforestry: introduction to, 169; around the world, 169–72; for humus formation, 73; in Nepal, 167–68; rain forest regeneration, 173–75

agro-industrial complex (Big Ag), *see* Big Ag

Aktion Zivilcourage (Action Civil Courage), 99, 101

algae, 7, 15, 199

All My Cycle (Matsuzaka), 189

Amazon cities: differences from European farming, 38; disbelief in, 34; forest gardens, 39, 43, 44; Spanish accounts of, 33–34; terra preta from, 37, 39–40, 41–43

anaerobic dry toilet, 191–92

Añorve, César, 182–84

antibiotics, 19

Appropriate Rural Technology Institute, 93–94

aquifer depletion, 17. *See also* water

Aspergillus niger, 44

Australia, 110–11, 188

Austria, *see* Kaindorf eco-region, Austria

Aztecs, 38, 179–80

bacteria, 7, 9–10, 19–20, 41, 96. *See also* microorganisms

banana leaves, 170

bananas, 28–29, 118

Barichara, Colombia, 175

Bartholomew, Mel, 164–65

BASF, 15

A Basic Exercise (Matsuzaka), 190

Bates, Albert: *The Biochar Solution,* 161

Bayer, 22, 28

beef production, 16–17. *See also* cattle; dairy farmers

beekeeping, urban, 163

DAVID SUZUKI INSTITUTE

The David Suzuki Institute is a non-profit organization founded in 2010 to stimulate debate and action on environmental issues. The Institute and the David Suzuki Foundation both work to advance awareness of environmental issues important to all Canadians.

We invite you to support the activities of the Institute. For more information please contact us at:

David Suzuki Institute
219–2211 West 4th Avenue
Vancouver, BC, Canada V6K 4S2
info@davidsuzukiinstitute.org
604-742-2899
www.davidsuzukiinstitute.org

Cheques can be made payable to The David Suzuki Institute.